PRAISE FOR *NO PLACE TO HIDE*

A broad-ranging view of the critical issues facing our planet written in an easy to understand non-technical manner.

— Jay Zwally, University of Maryland Earth Sciences and NASA Goddard Space Flight Centre

A great read. A massive amount of information well organised and expressed.

— Sir Alan Mark, Fellow Royal Society of NZ, Knight of the New Zealand Order of Merit

Wonderful – for the first time, I have read an unbiased overview of the science behind global warming.

— Rolf Dobelli, Author of *The Art of Thinking Clearly*

This is a splendid summary of the global warming problem, and I hope it receives the attention and acclaim it deserves.

— Merrill Hiscock, Professor, University of Houston

I would sooner expect a goat to succeed as a gardener,
than expect humans to become responsible stewards of the earth
James Lovelock, *Gaia*, 1991

Would a revolution reduce the population and check production?
Then certainly I want a revolution.
Lord Edward in Aldous Huxley, *Point counter point*, 1928

He looked up at the sun and shook his fist:
'You ruddy old thing, we'll get you yet.'
H. G. Wells, *Trapping the sun*, 1914

NO PLACE TO HIDE

CLIMATE CHANGE

A short introduction for New Zealanders

Jim Flynn

CONTENTS

Part B: CARBON AND OUR FUTURE

Part C: WHAT IS TO BE DONE?

ACKNOWLEDGMENTS

I want to express my gratitude to the scholars with whom I corresponded, all of whom read passages from the manuscript (a few read early drafts in full): Christopher Barnet, Ken Caldeira, Gideon Henderson, Aynsley Kellow, Alan Mark, Roger Pielke, Jr., Stephen Salter, Kevin Trenberth, and Jay Zwally. All mistakes are, of course, my own. I received detailed advice on presentation from Grace White and style suggestions from Emily Flynn, Hannah Steiner, and Wendy McGuiness.

Those who read it from the point of view of a general reader included Rolf Dobelli, Ramesh Thakur, Jeff Scott, and Gilberto Corbellini, who has done so much to promote my work in Italy. Above all, I want to thank William van der Vliet, computer programmer for the University of Otago Psychology Department, whose work on the figures was both tireless and creative. Acknowledgments for permission to use various images are in the text.

LIST OF FIGURES

PROLOGUE

FREEDOM AND URGENCY

Coming to terms with climate change means accepting two propositions:

- The carbon content of the atmosphere is crucial. All that is really up for debate is what consequences will occur at what levels.
- The conventional approach to climate change (negotiations to reduce carbon emissions and so forth) sets targets that are not low enough, with deadlines that are far too late. Therefore, we must review proposals to buy time (keep temperatures under control) and use that time to get really clean energy.

When I began researching this topic it was a matter of personal empowerment. I was assailed by contradictory opinions that ranged from nightmare scenarios to reassurance, and had no idea of whether I should relax or be alarmed. This was intolerable. The most daunting thing was that I did not know how to start. I felt overwhelmed with the sheer mass of material available and incredible range of topics that fall under 'the environment'. I gradually found my way and wrote this book. I think it will help you to isolate the key issues and the most relevant information. If you are like me and critically examine what you read and hear, you will not elevate it to the status of scripture but instead assess everything I say.

In my opinion, *both* sides in the climate debate are in denial. Alarmed colleagues had prepared me for the sceptics' position: that we can go on with business as usual and leave climate change to fate. But I soon discovered that, like the sceptics, many of the alarmed were also prey to a delusion: that there is some hope we can persuade nations to cut their carbon emissions. The first group may defy what science says about the real world. But the second group ignores political reality: the world's leaders will not curb economic growth if it jeopardises their survival. They will not commit political suicide simply because a cause is just.

There is no long-term solution without the use of clean energy. However, its widespread adoption is too far away to prevent things happening in the meantime that are irreversible. For example, over the next 20 years, perhaps even 10 years, I believe that the Greenland and West Antarctic glaciers will begin to shrink dramatically and rising sea levels will be waiting in the wings. Before this happens, we must do something to buy time until clean energy comes to our rescue.

How can we quickly stop temperature rise? There is only one proposal worth taking seriously: pumping sea spray into the sky to make clouds whiter so they reflect more solar energy back into space. At this point, the reader may be tempted to shut this book because you suspect that the author has been reading too much science fiction. In fact, the urgency is such that some of the best minds are moving in the same direction. The US National Science Foundation has given a grant of $US20 million to investigate this strategy. However, even if this or any technology provides us with a period of grace, we will need clean energy as quickly as possible: the best bet is laser or plasma fusion of heavy hydrogen.

Part A is about our climate, past and present

Chapters 1 and 2 sketch climate history over a period long enough to isolate the natural (unaffected by humans) causes of climate change. It concludes that these are unlikely to have a dramatic effect in the foreseeable future. Chapter 3 argues that what climate does to us is essentially in our hands. The crucial question is whether the polar glaciers are melting and why.

Part B outlines the carbon-climate relationship and paints a grim view of our future

Chapter 4 outlines basic carbon chemistry and examines the consequences of increasing the amount of CO_2 (carbon dioxide) in the atmosphere. Chapter 5 predicts that by 2100, sea levels may rise so high that many coastal cities will have to be abandoned. The future productivity of agricultural land will see some winners as well as losers. But what with population spiralling toward 11 billion, a continent like Africa and a subcontinent like India face starvation. It also discusses 'the point of no return': the case that we may soon drift past a deadline after which temperatures will rise for a very long time no matter what we do. Chapter 6 shows that the negotiations since the Kyoto Protocol (1997), and the policies the major states are likely to adopt individually, are futile. It details why so little has been done to date and why the recent Paris accords will provide no solution. The political elite will not risk compromising economic growth. That is why realistic climate scientists predict that we will pass the point of no return by about 2050.

Part C asks: What is to be done?

Chapter 7 discusses the hopes of those who believe that the technology that exists (nuclear power, natural gas, carbon capture) will save us. The possibility that economic growth will not survive the exhaustion of oil, coal, and natural gas is acknowledged but I argue that even if this occurs, it will do so only after the point of no return and merely add to our woes. Chapter 8 offers my own personal solutions. We must use ocean spray to halt the trend toward rising temperatures before we reach the point of no return. We must make hydrogen fusion commercially viable before the end of the century.

The Epilogue concludes that all of us, the very alarmed, the somewhat alarmed, and the sceptics, should be able to agree on a common programme, if we put politics and recrimination aside in favour of sympathy with one another's anxieties.

That last chapter is followed by a reading list of 26 books and articles that may help you to get started on your own research and enable you to tell me if I am wrong.

Note: All temperatures are in degrees Celsius. To convert to Fahrenheit, take the former and multiply it by 1.8. Thus, when a model predicts a 6-degree Celsius temperature rise by 2100, this translates into 10.8 degrees Fahrenheit (6 × 1.8 = 10.8).

— PART A —

THE HISTORY OF OUR CLIMATE

CHAPTER 1

CLIMATE HISTORY: DISTANT TIMES

The debate about whether we are doing something significant by putting carbon into the atmosphere is blurred by this fact: so many things affect Earth's climate it is difficult to isolate the possible effects of carbon from everything else. As the past is a guide to the future, the best tutorial about what these other things are is to review the history of Earth's climate and atmospheric carbon. In doing so, perhaps we can determine whether or not what nature has in store for us is more important than anything we are about to do to our climate.

Reconstructing History

Scientists reconstruct the recent history of Earth's temperatures using proxies, that is, things that correlate with how hot or cold it is at some time. Examples of proxies include tree rings, corals, layers of sediment, pollen fossils, ice cores, and bore holes.

Tree rings are normally wider during warm periods and narrower during cold ones and can provide temperature estimates over the last 1000 years. However, slow growth (narrow rings) can also be due to drought induced by warmer rather than colder temperatures. Indeed, there is evidence that tree rings have been poor proxies for recorded temperatures since 1960.

Some corals also have annual growth bands that yield yearly estimates of ocean temperatures. The corals form atolls whose top is dead and mostly above the water line, but whose perimeter is mostly submerged and alive. Analysing their record of growth gives estimates of sea-level change. As we will see later, these provide a basis for temperature estimates. Because corals seldom persist longer than a couple of centuries, only rarely do the growth rings go back more than 200 years.

Layers of sediments called 'varves' form on lake floors. Each year rivers carry sand, rock fragments, and soil to the lakes and these sink to the bottom. Lakes that contain little or no life-sustaining oxygen are the best sources of these sediments because burrowing animals cannot survive and thus will not disturb the layers of sediment. Summer temperature, winter snowfall, and annual rain determine the composition of the run-off into the lakes.

Pollen grains are produced by all flowering and cone-bearing plants and can be used to identify the type of plant from which they came. They too are preserved in the sediment layers in the bottom of a pond, lake, or ocean, so an analysis of the pollen grains in each layer tells us what kinds of plants were growing at the time the sediment was deposited. Since we know what plants can survive at what temperatures, the change in vegetation gives us a picture of changing temperature over time. Pollen fossils go back thousands of years, indeed, the weight of opinion is that they are reliable up to 70,000 years ago. The temperature tolerance of various species of plants is a far better proxy than the annual growth of trees.

Ice cores provide the best proxies. The deepest drills are informative for dates as early as 800,000 years ago. There are many problems of dating and estimation but ingenious solutions exist. For example, ice traps gas bubbles that are traces of the ancient atmosphere. There is a correlation between the average annual heavy hydrogen (deuterium) content of snow and the average annual surface temperature: the more trapped deuterium, the warmer the year. The deuterium content of ice also correlates with the level of CO_2 in the trapped bubbles, which provides a second measure. Ice core data on their own do not prove that atmospheric CO_2 causes climate change, however. It may be that temperature change alters CO_2 levels. Naturally, ice cores can tell us only about areas with polar ice or mountain glaciers.

Boreholes have the advantage that there are thousands of them all over the world and they go deep enough to measure thousands of years. The ones that yield useful data must be in areas where ground temperatures have been undisturbed by the presence of cities or farming. Also, underground springs or geological shifts must not have affected the borehole. Measurements must allow for the effect of heat rising upward from within the planet. Like most proxies, the further back in time, the less certain the measurements.

Many geologists believe that boreholes are the best proxy outside the polar regions, and that they may in fact give more plausible results than those generated by lumping all proxies together. How do their data approximate temperature? Ground surface temperature sends a thermal wave downwards at a known rate. Over time, each altered ground temperature sends down its own peculiar wave. As a wave descends, it leaves no effects behind it to contaminate subsequent waves, and it never catches up with the earlier waves below it. You find a hole, measure the temperature of the substrate (soil or rock) at intervals as you go down, and you find a record of every different temperature beginning with the present and going deeper into the past.

Imagine a series of elevators that go down a shaft, one floor per year for a thousand years. Each one contains its own reading of the temperature (its own thermal wave) at the moment it left. You are in a parallel shaft and can descend as deep as you wish in an instant. All you have to do is look inside the elevators one by one and copy the temperature readings therein.

The Distant Past

It is important to note that the proxies can be reliable as far back as 800,000 years ago but not much before. When we go further into the past, the force that swamps every other influence on climate is the movement of tectonic plates. These divide Earth's outermost shell (its crust and upper mantle) into chunks like the pieces of a jigsaw puzzle. There are eight major and many minor ones, some continental and some oceanic. Continental drift refers to the theory stating that these plates move about, sometimes colliding with one another, sometimes separating further and further apart. Therefore, over millions of years, they create and destroy continents as well as the ocean floor. We know that they caused massive volcanic eruptions that had a profound effect on Earth's atmospheric CO_2 and its temperature but that is about all we know.

One billion years ago, there was only one continent, Rhodenia, and none of its mass was near the poles. It broke up some 750 million years ago, although the continents may have come together briefly in the disputed supercontinent of Pannotia about 600 million years ago. By 550 million years ago, it had definitely disintegrated into four continents. The next supercontinent to form was Pangaea. It existed from 300 to 150 million years ago and also had no mass near the poles: what would become Antarctica and Australia were parallel and north of Australia today. It began to break up about 200 million years ago into Eurasia (which included the ancestors of North America and Eurasia) and Gondwanaland, which included South America, Africa, Australasia, and Antarctica.

The world we see today had pretty much assumed its present shape by about 56 million years ago – New Zealand was close to its present location, having started separating from Australia about 85 million years ago. At that time, there occurred a 5,000-year hot spell that was the closest parallel to current conditions. Baerbel Honisch of the Lamont-Doherty Earth Observatory presented the findings of 21 researchers investigating that period in 2012. Greatly increased volcanic activity had pumped CO_2 into the atmosphere at an unusual rate. During the 5,000 years, the atmospheric concentration of CO_2 doubled and the oceans responded by becoming more acidic. Looking at ancient mud under the ocean floor, they found that many corals and many single-celled organisms became extinct during that time, which is indirect evidence that other plants and animals higher on the food chain died out as well.

Ice Ages

Some 56 million years ago, polar ice caps did not exist. This was a rare state for Earth – polar ice caps normally persist for about 30 million years and can last for up to 300 million. But back then, the polar regions were temperate and covered with trees, rather like New Zealand or the Pacific Northwest of the USA today. Then, the planet began to cool, and about 30 million years ago, Antarctica became largely ice bound.

Accounts of climate history often use terms inconsistently, in particular 'ice age'. Even climate scientists use this term differently. Some use it to refer to a time when the polar glaciers advanced to cover much of Europe and North America. Others call any period during which polar glaciers exist at all an 'ice age' – and call the advance of ice 'glaciations'. I prefer the second usage, but throughout the book will choose language that leaves the reader in no doubt about what is being discussed. An 'interglacial' is the period during which the ice contracts (we now live in an interglacial). There are glacial 'maximums' (or 'peaks') and 'minimums', when the polar ice advances and retreats, respectively.

Figure 1 charts the last million years and gives information about glaciations during that period. We are sometimes told that the polar ice caps advance every 95,000 years (this is often rounded off to 100,000); and that interglacial periods are rather short, lasting only 10,000 years. However, if you take a ruler to Figure 1 (one mm = 10,000 years), you find that the peaks of glaciations come along anywhere from every 65,000 to 125,000 years, and that how long each one lasts varies hugely. The last interglacial interval lasted not 10,000 years but over five times that length, that is, starting 128,000

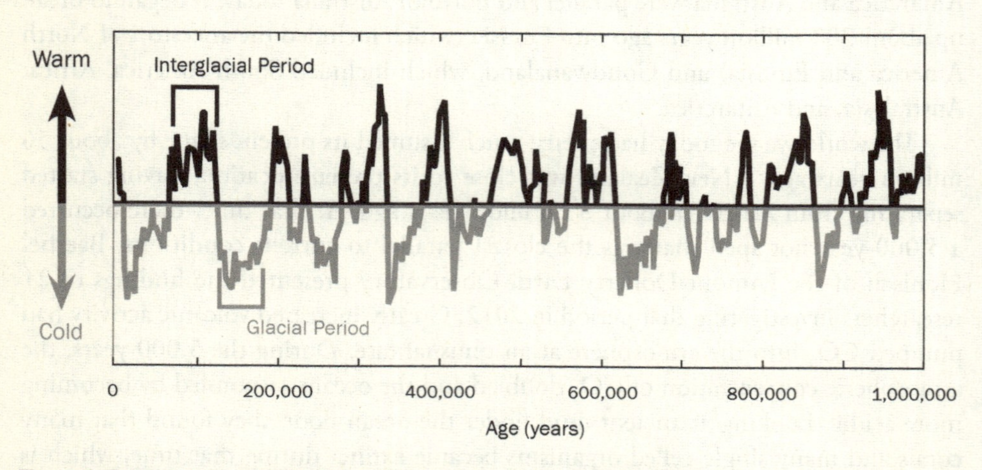

Figure 1. Temperatures for the last million years – from present to past.
http://www.divediscover.whoi.edu/iceage/timeline.html. THANKS TO Dive and Discover©Woods Hole Oceanographic Institution.

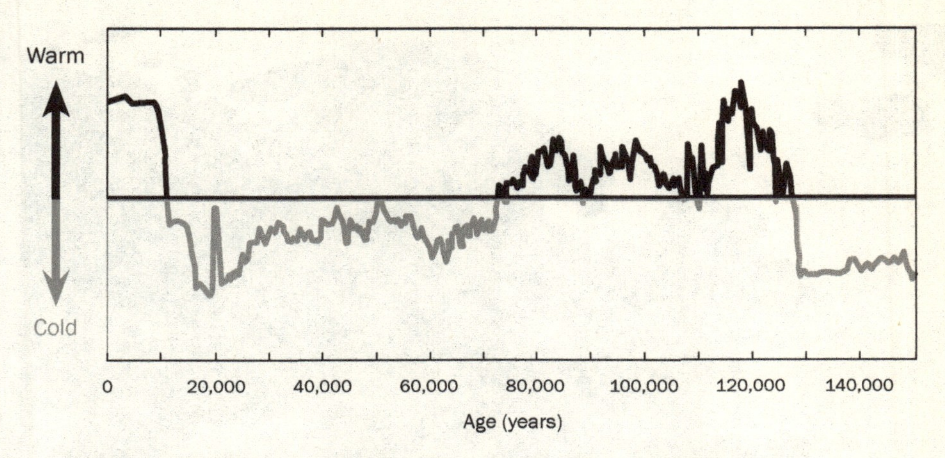

Figure 2. Temperatures for the last 150,000 years – from present to past.
http://www.divediscover.whoi.edu/iceage/timeline.html. THANKS TO Dive and Discover©Woods Hole Oceanographic Institution.

years ago and continuing until 74,000 years ago. Figure 2 shows what actually happened during that time (see the darkened section of the line). The entire period of 54,000 years is described as 'interglacial' but its temperatures were hardly constant.

The Wurm Glaciation

The subsequent, and last, glaciation was the Wurm Glaciation (the Otira Glaciation in New Zealand, and commonly referred to simply as 'the Ice Ages' the world over). Its temperatures are visible in Figure 2 as the grey line from 74,000 years ago until 12,000 years ago. It was at its peak about 20,000 years ago. As Figure 3 shows, glaciers (the black areas) once covered Europe almost down to the northern border of France, including most of Germany, all of Poland, and much of European Russia. In addition, they covered Canada and the USA's upper mid-west. Immediately south of the ice, spruce, fir, and arctic willow grew only in the more sheltered river valleys, and the rest of the landscape was mostly bare of forests. Human beings retreated south and learned to cope with the cold. They sewed clothes from animal hides, built shelters from mammoth bones, and buried food in permafrost to keep it from spoiling.

The map also shows the coastline at the time (edge of the pale grey). Sea level was 100 to 140 metres below what it is today, so areas of land were continuous that today are separate. For example, Ireland and Britain were part of the European continent. This marked change in landmass suggests unpleasant things are in store for us if today's glaciers were to disappear entirely.

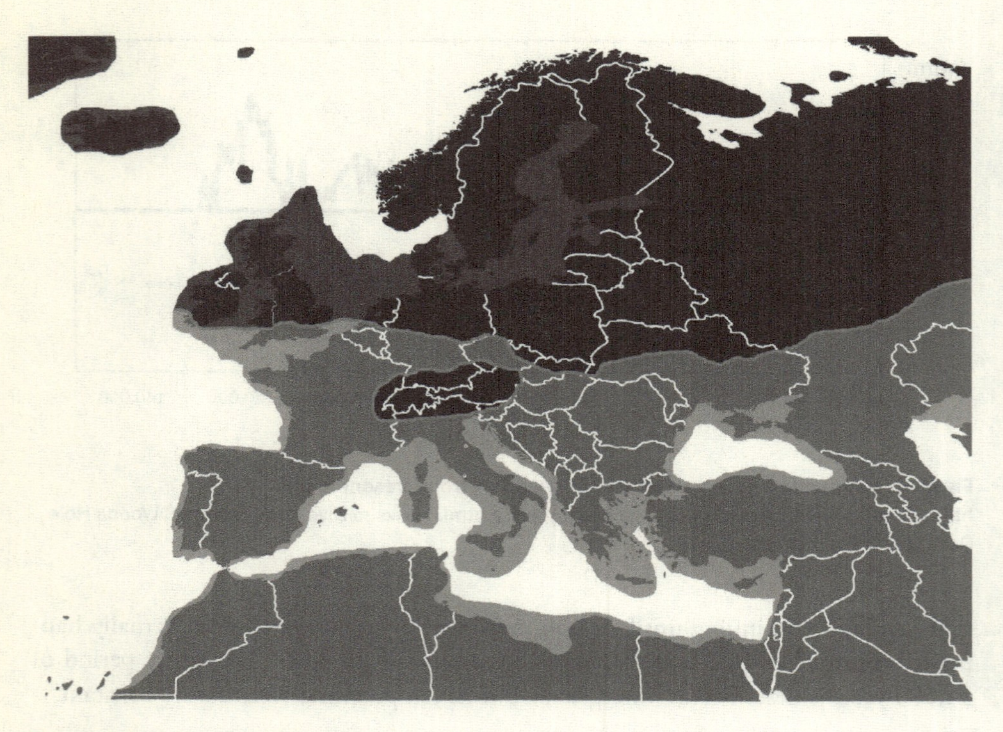

Figure 3. Wurm Glacial Maximum (Europe). Pale grey shows the area of exposed land at the time, dark grey is today's land mass and black is the Wurm Glacier.
http://commons.wikimedia.org/wiki/File:Ice_Age_Europe_map.png. AUTHOR Kentyne.

A blog called *Answers in Genesis* (AiG) advises us not to expect to read about this period in the Bible: '[the Bible's] focus is on events that took place in the Middle East after the Flood.' The blog is more scientific than most similar sources. It does not put creation at 6,000 years ago and does not deny that the Wurm Glaciation ever occurred. Whether we depend on science or biblical prophecy, it is good that the next glaciations seem remote. We have seen that they have serious consequences. But if adjusting to new glaciations would be a challenge, think what we would face if, thanks to carbon emissions, we began to heat the planet and eliminated polar ice entirely and sea levels rose accordingly.

CHAPTER 2

CLIMATE HISTORY: RECENT TIMES

We now pass on to relatively recent times during which we get firm dates about when people began to settle or abandon new areas because of climate change, and which even offer us written records of their observation. Figure 4 covers the last 11,000 years and shows that within our interglacial, there have been six hot and five cold 'snaps' – periods during which temperatures varied less than during glaciations or interglacial eras but enough to inconvenience humans.

These snaps vary from 200 years to 1,800 years in duration and should not be lumped together as causally identical. A minor cycle of Earth's orbit played a role 5,000–7,000 years ago. Its snaps are atypical, both in length and temperature pattern. They combined especially warm northern summers and cold tropical winters. Sometimes hot and cold snaps occur independently of changes in atmospheric CO_2. I will describe the most recent ones and then explain why. I should add that the data

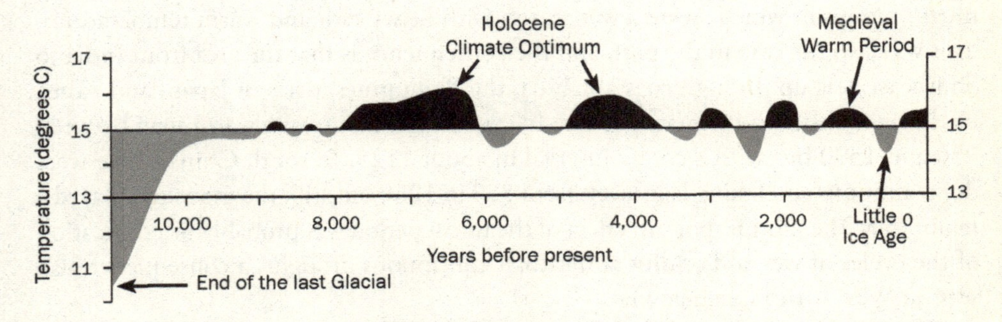

Figure 4. Temperatures for the last 11,000 years – from past to present.
THANKS TO Steve Gorman (2012), *The mad, mad world of climatism*, Chapter 4. Original data from Dansgaard and colleagues (1984).

are more complete for the Northern Hemisphere in general, and Europe in particular. Thus my narrative tells us less about some areas than we would prefer.

Global warming: The Medieval Warm Period

We tend to think of our recent warm temperatures as unique. But various parts of the world tell a different story. From 1100AD to 1400, there was a 'Medieval Warm Period'. This was followed by an abrupt transition to the 'Little Ice Age', which lasted about 450 years (from 1400 to 1850; see the next section, 'Global cooling'). There were regional variations. In Europe, the warm period was at its height from 1150 to 1300, and warm but unsettled conditions persisted until 1400. A bit earlier, from 1000 to 1200, there was unusual warmth in the Arctic, Canada, and Greenland. Earlier still, the Mayas in Mexico and Central America experienced warmth (and therefore drought) from 700 to 1200. No one knows enough to offer a complete explanation of such regional variations.

Australasia New Zealand had unusually warm periods from 1137AD to 1177 and from 1210 to 1260. They are thought to have been about as warm as today. If that is so, they hardly posed any problem of survival for the very first Maori (Polynesians who settled New Zealand around 1000). There are few data for Australia, which has been prone to drought over the whole of the last 1000 years. However, the Worimi Swamp of coastal New South Wales shows a pattern of more fires during the relevant 300-year period, which suggests unusually warm and dry conditions. The fact that Tasmania shows little impact from the Medieval Warm Period is not relevant because its weather differs from the mainland.

Asia The Tibetan Plateau had an extended dry period from 1075AD to 1375, which affected its nomadic inhabitants adversely. As they moved west, they started conquering cities, getting as far as Hungary. Elsewhere in Asia, except for the far north, monsoon winds create a wet season with heavy rain and warm temperatures. This was also the case in the past, and the consequence is that the area from India to Southeast Asia up through coastal China, the Philippines, parts of Japan, and Korea has its own distinctive climate pattern. In east China, there was a warm snap between 950 and 1300 but with a cooler interval in about 1125. In south China, there were dry conditions and falling lake levels from 880 to 1260 but still, the monsoon played a major role. The most important effect of the warm period was probably intensification of the cycles of wet and aridity rather than continuous drought, a consequence also seen in West Africa (see below).

The Americas During the Medieval Warm Period, parts of the present USA suffered from terrible droughts. Toward the east, the Hudson River Valley suffered from persistent drought between 800 and 1300. Toward the west was the Great Basin, covering a million square kilometres between the Rocky Mountains and California.

Its inhabitants depended on periodic wet years (these were not very wet) to forage for food. From 935AD to 1300, four droughts persisted for decades.

Conditions were even more severe in the Mojave Desert, which includes Southern California (Death Valley), Utah, Nevada, and Arizona. The Pueblo peoples of New Mexico lived in a semi-arid area and cultivated maize and beans with the help of an effective irrigation system. By 1100, they had constructed the monumental 'great houses' still to be seen in the cliffs of Chaco Canyon, and supported a local population of perhaps 2,200. A 50-year drought began in 1130 and water gradually disappeared and by its end, the Pueblos abandoned Chaco Canyon, and headed north. In 1276, a 23-year drought scattered them to the few areas still served by rivers and lakes.

The Mayan civilisation of Central America also collapsed under droughts, but these occurred from 760AD to 916, earlier than the Medieval Warm Period. Thanks to sinkholes that let people reach the water table and wetter years beginning about 1100, the Mayans of Northern Yucatan still flourished until, weakened by internal wars, they faced the Spanish in 1519. However, the Medieval Warm Period affected all of tropical South America. The Chimu civilisation of the Andes managed to survive droughts between 1245 and 1310. But that part of the continent belongs to a different weather system and had centuries of cold (though dry) climate while Europe was warm.

Europe and the steppes Between 1100AD and 1300, Europe north of the Alps had bountiful harvests. Huge swathes of forest were cleared for farming, and new land was put into production at high altitudes. Grape production moved 500 kilometres north. In 1120, William of Malmesbury, travelling through Gloucester (southwest England), remarked on the fact that grapes were planted in the open, unsheltered by walls. In fact, England exported wine to France, stealing much of the French market. Grape-growing moved north on the continent into Prussia and southern Norway. Fruit trees abounded in England and grew wild. Fishing improved in the North Sea and herring, carp (which could now be raised in fish farms), and cod supplemented the diet.

The best index of how the various nations of Europe benefitted from the Medieval Warm Period is to compare their populations in 1000AD and 1340 (just before the Black Death). The combined population of Britain, France, the Low Countries, Germany, and Scandinavia went from 12 to 35.5 million. Italy's population doubled, but semi-arid nations such as Spain, Portugal, and Greece made no gains, perhaps because the heat brought drought. The populations of Russia, Poland, Lithuania, and Hungary only rose by 35 per cent but, as we shall see, they bore the brunt of the Mongol invasions. Overall, the population of Europe as a whole went from 38.5 to 73.5 million during this period. The population explosion created cities and a new urban culture whose enhanced sophistication was a permanent asset: the Medieval

Warm Period made a huge contribution to the dominance of Europe and the history of the modern world.

The effects were not all beneficial. The steppes of southern Eurasia extend from Hungary to Mongolia. The nomads of the Tibetan Plateau were entirely dependent on the horse for war, travel, milk, cheese, and hides. When heat and drought dried up the grass – the horse's fodder, the nomads were forced to seek more fertile pastures and headed west, discovering cites that were easy to loot. By 1300, Genghis Kahn and his sons had obliterated the Turkish kingdoms of central Asia, destroyed almost all of the cities of Eastern Europe, and made the Russian states into vassals.

Greenland and the Arctic Scandinavia shared the population growth of northern Europe during the Medieval Warm Period. In the Canadian Arctic and Greenland, the peak of the warm period was earlier, extending from 1000AD to 1200. In about 985, Eric the Red, exiled from Iceland (which had been settled in 900), pioneered two settlements on the west coast of Greenland. The 'Western' settlement was further north than the 'Eastern' settlement: 950 kilometres above the southern tip as opposed to about 100 kilometres. The Norse fished for cod, killed seals and walrus, grew hay and barley, imported cattle and timber (from Labrador), and sent ivory (obtained from the Inuit) back to Norway.

After 1250, temperatures began to slide and by 1300, they had fallen dramatically. The winters between 1340 and 1360 drove the Norse out of the Western Settlement and the Eastern Settlement was gone by 1450. There is some mystery about the fate of the last inhabitants, although we know that some intermarried with the Inuit (the latter have 5 per cent Nordic genes).

Africa From Spain and Algeria to Syria and Israel, the Medieval Warm Period prevailed. After 1040AD, the levels of all the lakes in East Africa's rift system fell, including Lake Naivasha in Kenya; Lake Victoria, bordered by Kenya, Uganda, and Tanzania; Lake Tanganyika, bordered mainly by Tanzania and the Congo; and Lake Malawi bordered by Tanzania, Mozambique, and Malawi. In East Africa, from 1180 to 1350, the Nile's overflow was very low and in 1220, there was mass starvation in Egypt.

In West Africa, there were a large number of droughts during the warm period, but its most important effect may have been to cause rapidly alternating dry and wet years. In the Niger River delta, which dominates much of this area, Mande culture was organised around adapting to drought and flood. We cannot compare West Africa with Equatorial Africa or southern Africa because they offer no good data.

There is little information available about the sea-level rise that was associated with the Medieval Warm Period. We do know that the North Sea rose by 60 to 80 centimetres, causing large areas of coastal Denmark and Germany to fall below sea level and floods that drowned 100,000 to 400,000 people. Flooding in Holland created the great inland lake called the 'Zuider Zee'. In England, there was a tidal inlet that reached 25 kilometres inland to Norwich, and Beccles (now almost 10 kilometres from the North Sea) was an important port.

This contrasts with the Pacific Basin. Patrick Nunn points out, 'there is no evidence that [the] sea-level change brought about societal change' for the islands of the Pacific. He makes specific mention of the Kuriles (near Japan), Palau, Yule Island, Tutuila (American Samoa), Rangiora (Cook Islands), Niue, Lord Howe Island, and the Galapagos Islands. I had thought that the rising sea levels of the Medieval Warm Period would have menaced people on the Pacific atolls, but apparently not. Of course, this does not preclude the ill effects that a warm period hotter than the Medieval Warm Period might bring, assuming that sea levels reach record highs.

The Transition: 1300–1400

I have said that the transition from the Medieval Warm Period to the Little Ice Age was abrupt. As usual, we have our best data from Europe. The transition is usually dated at about 1400. However during the preceding century, European weather became variable with warm and cold alternating over cycles of a few months, or 1 year, or 7 years, or a decade. The unprecedented millions of European inhabitants were at risk if the food supply failed. In 1315, Europe had an unusually wet spring and summer, which was followed by a cold autumn. Heavy rains drowned newly planted crops and much of what survived was killed off by the cold. Storage was primitive and could offset only one bad harvest at most. The rains came again in 1316 and the following winter was bitterly cold causing famine everywhere north of Austria. People began eating grass, cats, dogs, and pigeon dung. Europe's population fell by over 10 per cent from about 82 to 73.5 million and it was 1322 before normal harvests resumed.

The great rains affected the fertility of the soil for years to come, for example, erosion destroyed half of the arable land in wide areas of Yorkshire. Then came the Black Death. Between 1348 and 1485 there were 31 outbreaks in England alone and by 1351, Europe's population dropped from 73.5 to 50 million. Unsettled weather and bad harvests in 1351 and 1352 added to the carnage. Thankfully, the rest of the century gave Europe reasonably good weather.

In 1976, Michael Jim Salinger offered estimates of New Zealand temperature change since 1300. All estimates until 1860 are based primarily on the advance and retreat of New Zealand's glaciers. By 1350, Maori began to arrive in large numbers. They had to adjust to cooler temperatures than they knew in the tropics. Although temperatures may not have been as high as during New Zealand's Medieval Warm Period, they were relatively benign and lasted throughout the entire transition period in the Northern Hemisphere and after. By 1500, Maori had evolved the 'classic' culture that confronted Europeans when they arrived. However, Maori did not escape their own 'Little Ice Age': it simply came a bit later and lasted from 1600 to 1800.

As Helen Leach has pointed out, cold drove population and horticulture (mainly kumara) on the South Island toward the north. During his periodic visits between 1769 and 1977, Captain Cook found no Maori gardens as far north as Marlborough (near the top of the South Island).

Global cooling: The Little Ice Age

Brian Fagan's *The great warming* (2008) and *The Little Ice Age* (2002) give excellent accounts. He made good use of Jean Grove's book, also called *The Little Ice Age*, published earlier, in 1988, and Peter deMenocal's seminal article *Cultural responses to climate change during the late Holocene* (2001). I rely heavily on these sources. However, research moves so quickly that some updating was necessary. In particular, over the last decade or two, we have learned a good deal more about the Little Ice Age.

Europe After 1400, winters became harsher and longer, which reduced the growing season. Oddly, some summers were unusually hot. Between 1400 and 1500, wet weather brought below-average harvests about once a decade. Most people were still peasants practising subsistence agriculture, and food supply limited population increase. After 1560, cold temperatures and their accompanying storms had a profound effect. Glaciers in the Alps and Pyrenees advanced: they reached a peak about 1616 and did not begin to recede much until 1850. Throughout Europe, poor harvests and food shortages were endemic until 1624.

During the 17th century, Europe began a slow transition to scientific agriculture that led to population growth with food security, despite the cooler climate. Still, the years from 1600 to 1750 were harsh. In Scotland, two-thirds of highland farms were abandoned. Norway began to do less farming and more ship building. The cold of 1695 cost Finland 30 per cent and Estonia 20 per cent of their populations. Vineyards in northern France were abandoned and, between 1739 and 1740, ice covered all major northern European rivers and Holland's Zuider Zee. Peasants froze to death. Ireland lost its potatoes in the cold ground as well as 10 per cent of its people. Herring left Norwegian waters for those of England and the Netherlands further south, and cod disappeared from the North Sea entirely. Fortunately, from about 1450, Basques had been catching cod in the Grand Banks off the east coast of Canada (Newfoundland) and other European nations had joined them in the 16th century. Thus, the shortfall could be made good.

The riddle of the Dutch Beginning in 1600, the Dutch expanded their dykes and reclaimed land from the North Sea. I had always assumed that the Little Ice Age came to their aid and that colder temperatures lowered sea levels. However, Holland may have experienced sea-level rise throughout the Little Ice Age. Good records at Amsterdam date only from 1700: they show between 1750 and 1770 a rise of 5 millimetres. What was happening?

About 20,000 years ago, during the Wurm Glaciation, the glaciers rested on a fulcrum that ran roughly through Denmark. Although the glaciers extended south through Holland, that still put their centre of gravity to the north. Their enormous weight depressed the land north of the fulcrum (Scandinavia). As that side of the seesaw went down, the land south of the fulcrum went up (Holland and Germany). It was as if an adult sat on one end of a seesaw and a child on the other. Naturally, when the glaciers retreated, the effects of the seesaw reversed.

Therefore, for thousands of years, Scandinavia has been rising out of the sea (and experiencing lower sea levels locally) and the Netherlands has been sinking into the sea (higher sea levels locally). When the cold of the Little Ice Age came along, it did tend to lower sea levels globally. But all it did for Holland was to blunt the long-term tendency of Dutch sea levels to rise. Now that global temperatures are again increasing, the 'fulcrum effect' means that Dutch sea levels are rising faster than those of the world in general. From 1700 to 2000, they rose by 270 mm.

Outside Europe Data are fragmentary but generally, cold and drought prevailed, although varying around the globe. The Spanish retreated from South Carolina to Florida. The British who landed at Roanoke in North Carolina in 1587 had disappeared by 1591. Those who landed at Jamestown in Virginia barely survived, with 80 per cent of them dying between 1607 and 1632. The Chesapeake Bay area (bordering Maryland and Virginia) had a cold snap centred on 1600. Yucatan (Mexico) cooled beginning in 1400. Chile is ambiguous: Lake Puyehue shows a wet period from 1490 to 1700; but further north, the San Rafael Glacier did not advance until about 1750. In New Zealand, the Franz Josef Glacier advanced, almost reaching the sea in the early 1800s. Between 1500 and 1800, southeastern Africa was 1 degree colder than at present, although this period was preceded by 50 years of warmth. Australia had its coolest century in recent times between 1600 and 1700.

In other words, Europe's climate is not the world's climate. It is arguable that only the years between 1590 and 1610 were very cold everywhere. This does not mean that suffering was localised. China, Korea, and Japan suffered more in the 17th century than Europe. Drought and flood alternated and famines were severe.

Transition to Better Times

After 1850, the world switched away from cold toward the current warm period. In Holland, the cold ended early, around 1825. Norwegian, Alpine, and New Guinea glaciers (the Carstensz was the last) were in retreat by 1850.

However, there were the usual regional variations. As indicated above, Himalayan glaciers did not retreat until 1880. Cold and crop failure plagued Belgium and Finland as late as 1867, and brought famine to India and China in 1876. England had a cold decade beginning in 1879, and Switzerland's cold ran from 1887 to 1890. In

1890, one of Iceland's main glaciers actually surged forward 10 kilometres. In North America, New England and the Chesapeake Bay were still cold in the 19th century although the west was relatively warm. It was 1900 before the new warmth became universal. Since 1910, Australia has experienced temperature rises a bit higher than the global average and New Zealand has been fairly close to Australia's pattern except for an odd period of cold between 1900 and 1935.

Where We Are

I had never realised the extent to which Earth's climate history had affected humanity. Our survey puts the present into perspective. Thanks to the present state of Earth's plates, we have sizable ice caps over the poles. Therefore, it is not too hot. We live in an interglacial era that has persisted ever since the Wurm/Otiran Glaciers retreated. Therefore, it is not too cold. The Little Ice Age is over and we are basking in what looks like the best of all possible worlds. As we have seen, even relatively mild climate change can make things very unpleasant. If we were not meddling with our climate, we might enjoy another 10,000 to 50,000 years before the next ice age (30,000 years is the best bet).

It may seem odd to congratulate our era because of its polar ice caps. In fact, they are just about the right size and whether or not they persist will determine our whole future.

CHAPTER 3

TODAY AND TODAY'S GLACIERS

In addition to Earth's various small to moderate-sized glaciers, three great glaciers exist. One covers most of Greenland. It is the main part of the Arctic Glacier that extends west from Greenland across northern Canada, and east from Greenland across northern Norway, Russia, and Alaska. Not quite as large is the West Antarctic Glacier, but the title of the world's largest glacier goes to the enormous East Antarctic Glacier. It is 27 million cubic kilometres in volume and averages 3 kilometres thick or one-third of the height of Mt Everest (above sea level). The masses of the three great glaciers say it all – that of the East Antarctic ice is at least 22,500,000 gigatons (a gigaton = 1 billion tons), that of the Greenland glacier 2,700,000 gigatons, and that of the West Antarctic 2,100,000 gigatons.

Are Polar Glaciers Melting?

I will argue that the polar glaciers are melting. If this is not true, we would have every right to suspect measurements that tell us global temperatures are rising.

Sea ice is ocean water that freezes and therefore contains some salt. *Land ice* is a glacier that contains no salt and rests on a sizable area of land like Greenland or Antarctica. The volume of Earth's sea ice is tiny in comparison to that of its land ice. The latter is far more important in the long run not just because there is more of it; if it completely melts or slides into the sea, it will cause sea levels to rise. Conversely, if sea ice melts, it will not raise sea levels because it is floating on the ocean's surface and already affects ocean levels. (The ice cubes floating in your drink raise the level in your glass almost as much as they would if they were to melt into liquid.)

We can monitor the loss of land ice thanks to the GRACE satellites (the Gravity Recovery and Climate Experiment satellites the National Space Agency, NASA, keeps aloft). Using the distance between the GRACE satellites as a measure, the precise strength of gravity at a point on Earth's surface can be determined, and from that, the amount of ice – less gravitational pull indicates that there is less ice. There used to be various measures and they were not coordinated until 2012. Still there is reasonable information dating back to 1990. The NASA data show that the trend for total land ice over the last 15 years is clear. In 2002, a net gain of 2,400 gigatons; in 2008, no gain or loss; from mid-2014 to mid-2015, a net loss of 3,000 gigatons; since then, an extra 4,000 gigatons. That is 7,000 gigatons (billion tons) in just 2 years.

The loss of land ice lags behind Earth's increasing warmth because the glaciers are so huge they create their own, very cold microclimate at the poles. If we focus specifically on the great glaciers, the Greenland Glacier did not begin to lose mass until 2008. In fact, the huge East Antarctic Glacier shows small gains to this day. The latter is atypical, however, because of its location. The land on which it rests elevates it well above sea level. This exposes it to warmer air but not to warmer water: air melts ice more slowly than would water of the same temperature. Furthermore, warmer air can increase snowfall and add to ice mass, thereby outweighing any tendency of this massive ice sheet to decline. The West Antarctic Glacier is different. It is far more exposed to warm water. The land on which it stands is actually below sea level. In addition, an intensification of the powerful winds that encircle Antarctica means that they are drawing warm water from the ocean depths to the surface. Exposure to warm water is irreversibly melting the Thwaites Glacier. It is only one of several component glaciers of the West Antarctic Glacier. But it is the one that is closest to the sea and it restrains four inland glaciers from 'sliding' toward the sea.

I have pointed out that the impact of melting sea ice is dwarfed by that of melting land ice. However, it is worth looking at recent global trends for sea ice because some climate-change sceptics cite them as grounds for complacency. It is indeed true that the sea ice around Antarctica has increased by about 1.9 per cent per decade since 1985. You might think that this is the result of global cooling, but it is really the result of the melting of the West Antarctic Glacier, discussed above. When the cold, fresh water of the glacial melt enters the salty ocean, the new mix of water (with a higher proportion of fresh water than salt water) is more likely to freeze; in 2007, Jinlun Zhang showed that it thereby contributes to the sea ice. In addition, the Antarctic Circumpolar Current (ACC) that surrounds Antarctica plays an important role. The cold waters of this, Earth's strongest current, do much to keep warm ocean waters at a distance. Finally, humanity is putting less ozone into the atmosphere and for complex reasons this increases the velocity of the winds near Antarctica, according to recent research (from 2002 to 2009). It has been posited that a shift in winds is blowing ice away from the coast, thus allowing exposed water in some areas to freeze more readily during winter.

The situation in the Arctic, however, is different, and is consistent with recent warming. Since the beginning of continuous space-based observation in 1979, Arctic sea ice has been declining by about 3 per cent per decade. There are yearly variations thanks to special conditions. For example, the area covered by Arctic sea ice in 2013 was greater than that of 2012. But the trend is clear. According to the National Snow and Ice Data Center, from 2014 to 2016, sea ice began to decline again and 2016 set a record low.

Squandering Our Luck

I have emphasised how fortunate we are that polar ice caps have dominated our time on Earth. We must thank the position of today's tectonic plates. They affect the distribution of Earth's land mass and thus determine whether or not there are polar continents to support polar ice. Indeed, the distribution of the continents is ideal to a degree unmatched in Earth's history. We have a huge landmass (Antarctica) over the South Pole and this land isolates the point at which coldness is at its height. Greenland gives us a sizable mass near the North Pole.

The point is that the polar ice caps are not just a thermometer that measure rising temperature. They do much to control Earth's temperature. Their glassy surface reflects the sun's rays back into space. If they disappear, their demise will be so potent that whatever we do about the CO_2 in our atmosphere will become a sideshow. Global temperature rise will be out of human control.

Some sceptics accept this, but argue that the rising temperatures that promise to trigger the glaciers' collapse has nothing to do with atmospheric CO_2. Thus far we have ignored other possible causes of temperature rise.

Earth's Orbit and the Climate

Milutin Milanković was a Serbian mathematician, astronomer, and climatologist of the early 20th century. He studied the position of heavenly bodies in conjunction with the advent of glaciations. The advent of glaciations has to do with Earth's position in regard to the sun. The text below combines his insights with recent discoveries by NASA. Over time, changes in Earth's gravitational interaction with Jupiter and Saturn alter Earth's orbit in three ways:

1. The shape of Earth's orbit varies from almost circular to more of an ellipse, which is like an elongated circle. A more elliptical orbit increases the distance between the sun and Earth. This cycle takes from 90,000 to 100,000 years.
2. The 'tilt' of Earth's axis varies. Imagine a table at one end of which you have embedded the sun. Earth spins on that table like a top. Although the top's axis would theoretically be straight up and down, that never quite happens. As the angle of Earth's axis comes *closer* to the perpendicular in relation to the sun, the summer becomes cooler and allows ice to persist, resulting in enhanced polar ice. This cycle takes about 40,000 years.

3. Earth never revolves steadily on it axis but 'wobbles' as it spins (look at your top again – it will be wobbling). This too means cooler summers and the cycle is about 23,000 years. NASA has recently discovered other causes of wobbling that until then had been obscure. For example, loss or expansion of polar ice (due to climate) affects Earth's mantle because it redistributes its mass. Also, the amount of water on the huge continent of Eurasia, whether the continent has lost water due to depleted ground water or drought (again due to climate), also affects the distribution of Earth's mass. This is a lesser factor compared to the size of the polar ice sheets.

The cash value of this is that astronomers think they can estimate the next glacial advance. They say it will come in about 30,000 years and posit a cycle of roughly 95,00 years. However, it is disturbing that this has been true only over the last million years, so the predictions are something of a guess.

At any rate, there is nothing here that threatens our immediate future or explains our immediate past. The models show that recent changes in Earth's orbit would be insufficient in themselves to cause the degree of warming that has occurred; indeed, some refer to any changes from these interactions alone as 'tiny'. I should note that there is the energy the sun emits, which has increased by 30 per cent over Earth's history (4.7 billion years). Clearly that factor will do little to influence the next few hundred years.

Ocean Currents and the Climate

If we can discount the heavens, how could we have gone through a Medieval Warm Period about 700 years ago followed by a Little Ice Age that ended less than 200 years ago? How could we go from a warm snap to a cold snap so quickly? The transition took little more than a decade or so. Robert B. Gagosian, Director of the Woods Hole Oceanographic Institution, argues that altered ocean currents caused these sudden changes.

Figure 5 shows a huge ocean current system called the 'Ocean Conveyor'. Its white segment is a current near the surface of the oceans. From the southern tip of Africa, it conveys warm waters northward that reach as far as the North Atlantic. There, the ocean gives up its heat to the atmosphere and prevailing winds carry the heat eastward, warming Europe.

The engine of the Ocean Conveyor is a 'vacuum' in the North Atlantic Ocean, which pulls the warm surface waters northward. This void is created when cold water sinks deep and then warm waters rush in to fill it. The cold waters sink only if they are very salty; it is salt that makes them heavy. Therefore, if fresh water from the melting Arctic Glacier dilutes the salt content of the cold water in the North Atlantic, the

Figure 5. The Ocean Conveyor. This system transports heat throughout the planet. White represents a warm surface current. Black represents a deep cold current. https://www.whoi.edu/main/topic/ocean-conveyor. Illustration by Jayne Doucette. ©Woods Hole Oceanographic Institution.

cold water does not sink; and no warm water rushes in to fill the vacuum. The whole system 'shuts down' and a cold snap (like the Little Ice Age) ensues.

The portion of the Ocean Conveyor that passes through the sea near Florida is called the Gulf Stream. Measuring its flow is a good measure of whether the Conveyor is in full throttle (keeping Europe warm) or weakening (threatening a cold snap). David Lund and his associates have done a fascinating study of the Gulf Stream. Their measurements show that its flow is high and its average temperature about 'medium'.

Volcanoes and the Climate

Volcanoes have a brief cooling effect on temperatures by creating a dust veil that cools by shielding us from the sun. The veil disperses to almost nothing by the fourth year from the time of the eruption. Ramond Bradley and Philip Jones have tracked the Dust Veil Index (DVI) from 1500 to 1990 for the Northern Hemisphere. The DVI predicts a sudden rise in temperature from about 1920 to 1960 (when the DVI was very low), followed by cooling thereafter. As we know, the Northern Hemisphere has undergone a temperature rise ever since 1850, with increases continuing to the present. There is simply no evidence that recent volcanic activity is responsible for the recent temperature rise.

Sunspots and the Climate

Sceptics who accept that global warming is underway but not that human activity is its cause often nominate sunspot activity as a cause of global warming. Sunspots are the result of intense magnetic activity on the sun that increases its brightness. However, to play the role of a major cause, sunspots would have to explain just how, over the last 800,000 years, sunspots also caused fluctuations in atmospheric CO_2 that match temperatures so well. I have never seen this addressed.

Sophisticated analysis of the evidence has begun to thin out the ranks of the sceptics. At one time, Richard Muller was a sceptic. He and his daughter founded the Berkeley Earth Surface Temperature Project, which has published five landmark papers (available online). They studied temperature trends over the last 250 years and, to his surprise, concluded that 'by far the best match was to the record of atmospheric carbon dioxide.' The hypothesis that solar variability (sunspots) was a cause of temperature change simply did not fit. This tallies with recent data that solar activity has little effect on the brightness of the sun. Small variations over a few years were attributable to volcanoes and ocean currents.

Muller now predicts a rate of temperature rise that would amount to 2 degrees by 2070. Sea temperatures would rise a bit more slowly, so his projection would amount to a 2-degree ocean temperature rise at about 2100. As we will see, I think that these are very conservative estimates. Muller does caution that if China continues to escalate its use of coal, we may have to wait no more than 10 or 20 years (until 2032) for the 2-degree increase.

Learning From History

What natural processes have caused past temperature changes, that is, what causes operated independent of the interference of humans? We know much about three levels: plate tectonics and the existence of polar ice; the planet's orbit and glaciations; ocean currents and cold/warm snaps. As for the present and near future, we can rule out the sun's total energy output, volcanoes, sunspots, and continental drift. Today, there is no plausible candidate as a cause for major climate change except ourselves. We have become a force so great that we can actually usurp the potency of plate tectonics and eliminate the polar glaciers without altering the distribution of the continents. Perhaps we should feel 'God-like' except that we are only half aware of what are doing and will find the consequences so unwelcome. The whole climate debate really turns on the effect of our contribution to atmospheric carbon, and carbon is the subject of the next chapter.

— PART B —

CARBON AND OUR FUTURE

CHAPTER 4

ALL ABOUT CARBON

When carbon reaches the atmosphere, one carbon atom combines with two oxygen atoms to make CO_2. As we will see, the primary carbon-based fuels are wood, coal, oil, and natural gas, and they differ in the amount of carbon they add to the atmosphere. For example, the main component of natural gas is methane and it has the chemical formula of CH_4, meaning one carbon atom to four hydrogen atoms. Its low carbon content makes methane the 'cleanest' of these fuels (the hydrogen does no harm). Methane is also a greenhouse gas but I have not analysed it separately because its carbon, like all carbon, forms CO_2 in the atmosphere almost immediately.

What is the Greenhouse Effect?

Earth's surface absorbs visible radiation (light) from the sun, which causes heating. At the same time, the surface and the atmosphere emit infrared radiation back to space, which produces cooling. Over a long period, surface temperature will remain constant because the amount of heat absorbed, as visible light, is equal to the amount of heat emitted as infrared. Nitrogen, oxygen, and argon together make up more than 99 per cent of the atmosphere. None of these three gases absorb either light or infrared and as far as temperature is concerned, it is as if the main components of our atmosphere do not exist.

CO_2 and water vapour are different. They may be a very small component of our atmosphere but whether their levels rise or fall, even by small amounts, plays a key role. While they do not inhibit the light that reaches Earth, they do inhibit the escape of heat as infrared into outer space. The carbon of CO_2 is bonded to oxygen, making two bonds: $O \leftarrow C \rightarrow O$. These bonds vibrate at a frequency that absorbs heat energy. How does this impact on global temperatures?

We can turn to an analogy to answer that. Why does the inside of your car get so hot from sunlight even on a cold day? Glass has vibrational frequencies very like CO_2.

Sunlight has a short wavelength and easily penetrates the glass windshield. When it warms objects inside the car, they emit heat as infrared radiation, which has a longer wavelength and the longest of these cannot penetrate glass. Thus heat is trapped within the car. Glass is unlike the atmosphere in that by enclosing a space, it literally keeps the warmed air inside so it cannot escape at all. The CO_2 and water vapour do not create a glass barrier, of course. But when they absorb the infrared 'attempting' to escape into space, the infrared is scattered in all directions. Much of it goes downward, causing the warming effect. This is called the 'greenhouse effect', although the phrase must not be taken too literally.

The CO_2 in the atmosphere is measured in *ppm* or parts per million. This is a percentage that refers to how much of the atmosphere's volume is taken up by CO_2. For example, 400 ppm means that for every million molecules in the atmosphere, 400 of them are molecules of CO_2. For global warming, the key measurement is the ppm of CO_2 in the atmosphere. However, in reading the literature, you will come across scientists who talk about weight rather than volume. For example, they may say we have put so many gigatons (billion tons) of carbon in the atmosphere in a given period (referring to its weight *before* it combines with oxygen). Or they may talk about the total weight or mass of CO_2 in the atmosphere (its total weight *after* it has combined with oxygen). See Box 1 for guidance on how to convert from one of these to the other.

Box 1: Conversions

Conversion: how much carbon released into the atmosphere creates how much CO_2? One oxygen atom is 1.33 times as heavy as one carbon atom. Therefore, for CO_2, one carbon atom (1.00) plus two oxygen atoms (2 × 1.33 = 2.67) = 3.67. When carbon hits the atmosphere, the CO_2 it creates weighs 3.67 times as much as carbon alone.

Conversion: what volume of CO_2 in the atmosphere equates with the total mass of carbon in the atmosphere? All you do is use 2.08 as a multiplier: if the atmosphere contains 400 ppm of CO_2, it contains 832 gigatons of carbon. This multiplier is dependent on the actual size of Earth's atmosphere.

The Crucial Question of Water Vapour

Water vapour is the gas phase of water and is relatively common. It is continuously generated by evaporation and removed from the atmosphere by condensation. However, the warmer the atmosphere becomes, the greater its capacity to hold water vapour: each degree of extra warmth adds 7 per cent to the atmosphere. Physics shows that doubling atmospheric CO_2 *alone* would raise temperatures only by 1 degree. The combination of CO_2 *plus* water vapour has three to five times the potency of CO_2 alone in generating warmth. Therefore, the capacity of CO_2 to enhance water vapour is crucial to how much carbon emissions will increase global warming. Sceptics have demanded hard evidence.

Kevin Trenberth (National Center for Atmospheric Research) and David Easterling (National Oceanic and Atmospheric Administration) measured the increase of water vapour over the USA at between 3 and 4 per cent since 1970, an increase of almost seven and a half trillion litres (nearly 2 trillion gallons); this increase is typical. They used measurements of specific and relative humidity in the atmosphere, which give the quantity of water vapour above a particular land mass. It is now quite clear that water vapour feedback occurs. Climate change journalist Justin Gillis details the emerging consensus, which simply settles the question of whether or not to take the models of temperature rise seriously. Note that it was only very recently that the sceptics were refuted on this point.

The Historical Record

Nevertheless, some sceptics will urge that they have not been refuted entirely. They believe that history is on their side. In order to get a complete record of atmospheric CO_2 and Earth's temperature, they go back 600 million years and conclude that the two bear no relation to one another; and when they look at the last 1,200 years, they see temperature fluctuations when the level of atmospheric CO_2 was constant. Over the next few pages, I will evaluate three periods: the last 600 million years, 1,200 years, and 800,000 years. I will demonstrate why I think the first is too long and the second too short. Let us be clear about our methodology. The mere fact that other things affect Earth's temperature does not rule out the possibility that the level of atmospheric CO_2 affects it. What we want to do is estimate what rising CO_2 does when no other factor could be blurring its effects.

The last 600 million years Sceptics use Figure 6 to bolster their case. Beyond 50 million years ago, temperature jumps all over the place on the one hand, while CO_2 varies from 240 ppm (recently) to 7500 ppm (the distant past) on the other. The correlation between the two looks to be nil. Worse still, warm temperatures coincide with relatively low carbon levels.

However, the CO_2 data are chancy. Going back 800,000 years, we have ice core data, which are highly reliable at least for polar regions. Going beyond that to 100 million years, we must use ocean sediments. By 450 million years ago, we must use complex models, for example, a geochemical model (GEOCARB III) that builds in

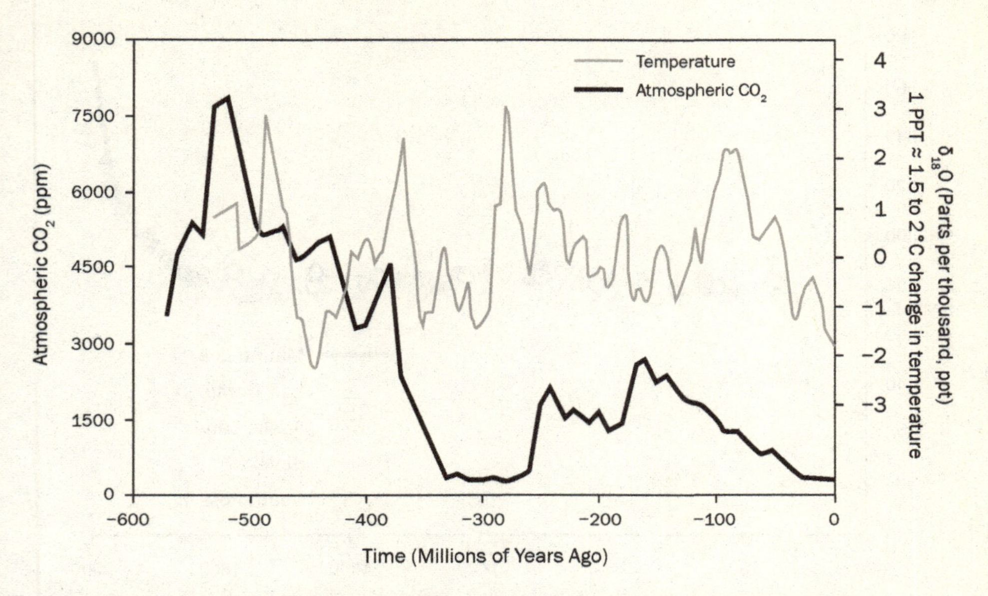

Figure 6. Atmospheric CO_2 and temperature over 600 million years. http://files.meetup.com/1429141/CLIMATE%20CHAGE_REPORT.pdf. THANKS TO Science and Public Policy Institute (SPPI).

volcanoes, erosion, and carbon deposits. Its results are largely independent of any direct evidence and can be taken as little more than first estimates. The levels of atmospheric CO_2 were probably higher in the distant past than today but no precision is possible. Setting aside the quality of the data, this immense period begins before the formation of the supercontinents and spans the subsequent history of continental drift. This is the Earth of Rhodenia and Pannotia and Pangaea and Gondwanaland, an Earth so different from today as to be almost an alien planet. The influence of the changing distribution of the continents on climate was so massive that it swamped every other factor.

The last 1,200 years Figure 7 covers 1,025 years before 1850, at which point humans began to put carbon into the atmosphere, and roughly 175 years since. It covers both the hottest time of the Medieval Warm Period and the coldest of the Little Ice Age, a period during which temperatures varied by about 1.6 degrees, at least in the Northern Hemisphere. And yet, until about 1850, atmospheric CO_2 was constant at about 280 ppm. What this reveals is that the Global Conveyor, including the Gulf Stream, and other minor factors *can* affect climate irrespective of CO_2 levels, a finding that climate-change sceptics have used to bolster their rejection of climate models. Nonetheless, we cannot use a period in which carbon does not vary to measure the impact of carbon when it does vary.

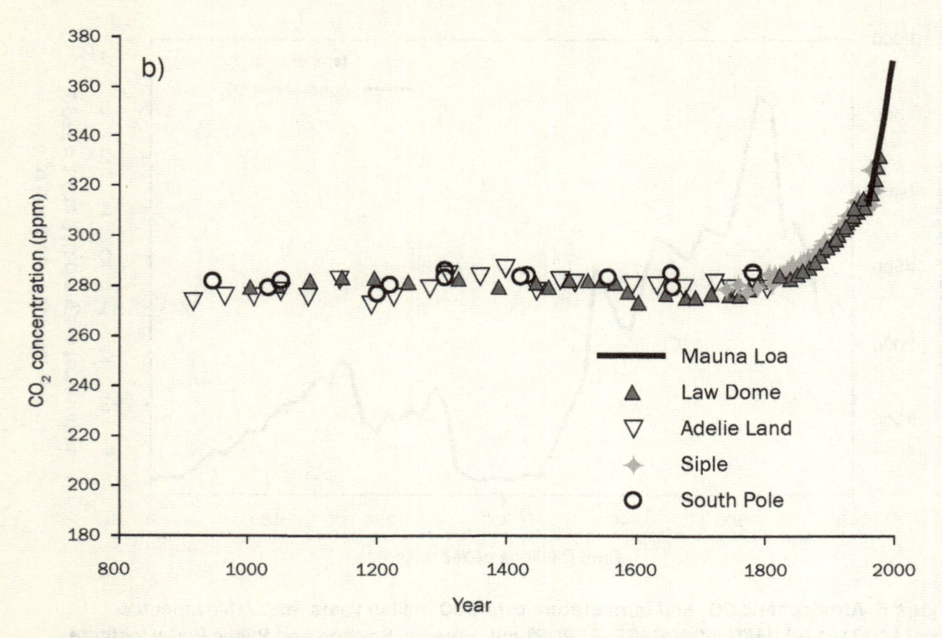

Figure 7. Atmospheric CO_2 over 1,200 years (from Antarctic ice cores and Mauna Loa). http://www.grida.no/publications/other/ipcc_tar/?src=/climate/ipcc_ tar/wg1/fig3-2.htm, image (b). From: *IPCC Third Assessment Report – Climate Change 2001 (Working Group I: The Scientific Basis).* Recent atmospheric measurements at Mauna Loa (Keeling and Whorf, 2000) are shown for comparison.

The Wrong Foot

It did not help that when orthodox scientists first reconstructed the last 1,200 years they did it badly. The initial reconstruction actually indicated that the Medieval Warm Period and the Little Ice Age had never happened. This hardly inspired confidence in the methodology of estimating the history of Earth's temperature.

To show what went wrong, and to vindicate the methodology I have followed in this book, I will compare some reconstructions of the past 1,200 years in the literature. Unfortunately, these reconstructions are all graphed on different scales. Therefore, it is necessary to have a common year or 'zero' point to make them truly comparable. Global temperature was never zero of course. It is a just a matter of putting a date at zero, so we can see how much various temperature reconstructions vary on a common scale. I selected 1919 as year 'zero', which has the effect of making all temperature reconstructions much the same after 1880. The reader may wonder at this. It is because every scholar began to 'estimate' temperatures in this later period using mostly the same actual records of temperature. These data began in the mid-19th century, although they become fully accurate only after 1950.

In Chapter 2, I presented reliable historical testimony. We know that beginning about 900AD, there was a trend of increasing temperatures that reached Greenland a bit early (coinciding with the arrival of the Vikings), that there was a plateau in temperature in Europe from about 1100 (when grapes began flourishing in England) to 1400 (when grapes began losing ground), and that there was cooling thereafter (people ice skating on the Thames).

We can use this information to assess how valid reconstructions of temperature fluctuations over the last 1,200 years really are. My own historical graph of temperature changes (broken line in Figure 8) reads as follows: a gradual rise starts in 900AD and reaches a peak of +1.1 degrees about 1150; stability from 1150 to 1350; a clear decline by 1400, stabilising at -0.5 degrees from 1600 to 1800; a gradual rise to 'zero' by 1917; and another peak at +1.1 degrees today. You will note that my reconstruction puts the difference between the peak temperature of the Medieval Warm Period (which probably equalled the same temperatures we have today) and the nadir of the Little Ice Age at 1.6 degrees. This is reasonable in that all reconstructions indicate a difference of about 1.6 degrees. Reconstructions in this book will focus mainly on the Northern Hemisphere since the best data come from there.

My historical reconstruction includes the essential features that all reconstructions must accommodate. They should all show a rise to a plateau (Medieval Warm Period), followed by a broad valley (Little Ice Age), and then a rise to the temperature we have today.

In 1999, Robert Mann and his colleagues published a reconstruction that had tremendous impact. It is the thin line in Figure 8. It looks like a 'hockey stick' with the blade (the right-hand end of the line) pointing up, and the name has stuck. The line indicates temperatures so constant before 1900 that the rise from 1950 to 2010 looks dramatic and unprecedented. Comparing it to the broken line, you can see that Mann's reconstruction *virtually denies* the existence of both the Medieval Warm Period and the Little Ice Age. Those who questioned Mann, like my former student Aynsley Kellow, were attacked as obscurantist. In fact, Mann's reconstruction is so far from the historical record I cannot see how anyone could have been happy with it.

This harsh assessment must be tempered in the light of a recent study. The thick line in Figure 8 is the 2011 reconstruction by Bo Christiansen and Fredrik Ljungqvist. They used a comprehensive set of proxy data to estimate Northern Hemisphere temperature changes. For estimates back to 1000AD, they used 11 sources, including four tree-ring sources, always dubious, and four ice-core sources, all from the ice cap extending into Greenland and Northern Canada. The ice-core data meant a strong polar bias, and recall that the Medieval Warm Period there ended far earlier than in Europe, which is what their reconstruction shows. Another source is from Chesapeake Bay sediments (and the pollen grains therein) and the final two are multi-proxy

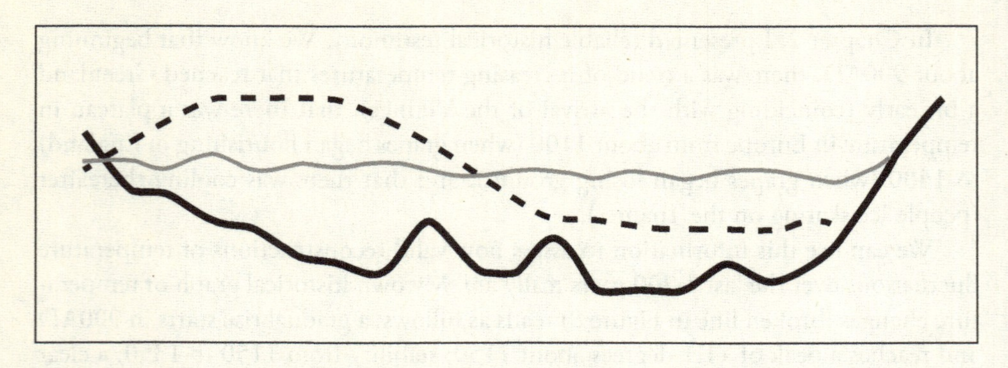

Figure 8. Reconstruction of Northern Hemisphere temperature trends from historical evidence (author's broken line), and from publications by Robert Mann and colleagues (1999; thin line), and from Bo Christiansen and Fredrik Ljungqvist (2011; thick line)

studies from China. Although their methodology appears painstaking, the polar bias means their results are almost as odd as Mann's. Their thick line captures the Little Ice Age (although it appears a bit too cold). However, their Medieval Warm Period peaks at about 1000 (far too early). After 1000, they suggest that temperatures fell precipitously, so much so that Europeans must have been hallucinating when they wrote about the balmy years of 1100 to 1300.

Figure 9 tells a different story. The broken line is once again my reconstruction. The thick line represents a reconstruction for the Northern Hemisphere published by Hubert Horace Lamb in 1990. He gave the Medieval Warm Period its name and charted what became called the Little Ice Age. His estimates of temperature change trends are based on scattered records about rainfall, flooding, winter temperatures, and an encyclopaedic study of historical accounts. Naturally, they shadow my historical reconstruction. However, his is probably better, showing detail that my crude yardstick lacks.

The thin line is Shaopeng Huang's more recent (2008) reconstruction of global temperature changes based on thousands of boreholes. It shows we are finally getting a reconstruction that both mirrors history and is based on scientific proxies. The Huang line registers the onset of the Medieval Warm Period. He has its temperature changes too low from 1150 AD to 1250 and they match history only by 1300. That year gives his highest estimate and it is a bit lower than the true historical peak from 1150 to 1300. His maximum estimate yields a temperature that would be about 0.5 degrees below today's temperatures, and I believe temperatures of the Medieval Warm Period were equal to those of the present day. On the other hand, he gets the Little Ice Age almost perfectly, although he has it in full force about 1650 rather than 1500. It is only fair to add that its date of onset is debated.

We must take into account that Huang's reconstruction is a global one, inclusive of both hemispheres, while my historical reconstruction is for the Northern Hemisphere only. As Chapter 2 indicates, the Medieval Warm Period and the Little Ice Age probably affected the Southern Hemisphere less than it did the Northern. This would explain why Huang's estimates tend to fall a bit below my peak and tend (on average) to be a bit higher than my lowest point. The important thing about his reconstruction is that both of these eras are *there*. Overall, the match is pretty good. Thus, he vindicates the methods science is using to reconstruct the past.

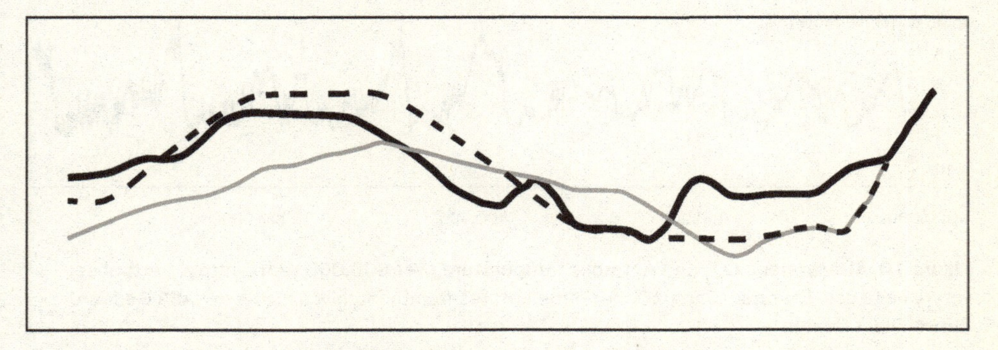

Figure 9. Reconstruction of Northern Hemisphere temperature trends from historical accounts (broken line) and from publications by Hubert Lamb (1990; thick line) and Shaopeng Huang and colleagues (thin line; 2008).

The last 800,000 years Figure 10 covers the last 800,000 years and the quality of the data is excellent. The CO_2 content of the atmosphere varies considerably, so we can at least look for a correlation between carbon levels and temperature.

No doubt, during this period there were warm and cold snaps but the dominance of carbon is so great that lesser temperature trends like this are lost in the data. Needless to say, the continents were stable over this time, so the huge interference of continental drift is gone. Only one main factor remains: variations in Earth's orbit. And, as I will show below, these variations confirm rather than blur the influence of carbon.

In Figure 10 the resemblance between the graph of Antarctic temperatures over time and the graph of atmospheric CO_2 over that same time is quite remarkable. However, there is a consistent short-time lag between the start of an upward temperature trend (as a colder glacial ends) and a rise in atmospheric CO_2. This seems to reverse cause and effect. The climate models interpret this as follows. Changes in Earth's orbit affect the energy the planet receives from the sun in terms of intensity and location, sometimes being responsible for beginning the cooling process, sometimes the warming. When the oceans became warmer, huge amounts of carbon

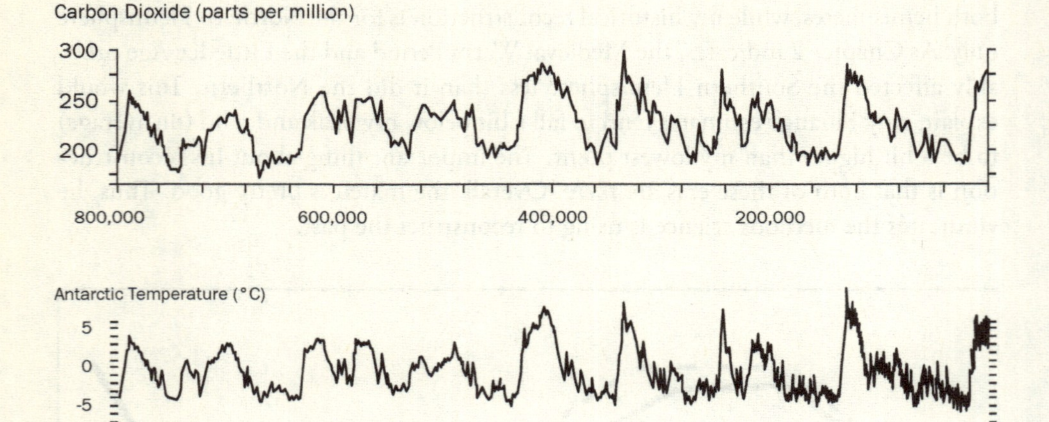

Figure 10. Atmospheric CO_2 and Antarctic temperature over 800,000 years. http://earthobservatory.nasa.gov/Features/CarbonCycle/?src=features-fromthearchives THANKS TO NASA Goddard Space Flight Center.

stored therein (and elsewhere) were released into the atmosphere as CO_2; this (plus water vapour) contributed powerfully to the warming, and Earth fully exited a glacial period and entered a warm interglacial period. Earth's orbit triggered a temperature rise but the carbon released finished the job. Today, we are the triggers of a massive carbon release, and carbon will have its way, no matter whether triggered by Earth's orbit or by ourselves.

As for the consequences of atmospheric CO_2, Figure 10 shows that temperature varied over a range of 7 degrees (averaging the peaks of glacials and the depths of interglacials). The corresponding CO_2 levels varied from 200 to 280. Therefore, an increase in CO_2 levels of 40 per cent (80 divided by 200 is 0.40) caused a 7-degree temperature rise.

A Direct Intervention

Scientists are never entirely happy with correlations established by history. The best test of a factor's effects is always to manipulate it under laboratory conditions. Ideally, we would radically increase CO_2 in a time period too short for anything else to count. Since 1850, humanity has undertaken just that experiment: pumped large amounts of carbon into the atmosphere during a mere 166 years. And we are beginning to get some preliminary results. I believe that the signs have been worrying for over 50 years.

Figure 11 gives data from this period. The rising curves of humanity's CO_2

emissions, the CO_2 concentrations in the atmosphere, and global temperatures give every appearance of a causal link. Between 1960 and 2008, atmospheric CO_2 increased from 315 to 375 ppm, which was a 19 per cent increase, and it appears that this raised temperatures by 0.60 degrees.

However, the figure does show exceptions to the rises going in tandem. Note that the temperature curve was relatively flat for the years from 1960 to 1978 and rose after 1989. Russia and Eastern Europe used to burn very dirty coal, which can be described as a high carbon–high sulphur fuel. Atmospheric sulphur acts as a sunshield and offsets the global warming tendencies of CO_2 and the combination has resulted in relatively stable temperatures. With the advent of clean air acts in the West and the fall of communism in the East, sulphur emissions declined and temperatures took off. The fact the Chinese are now burning more sulphurous coal may be slowing warming. But they are beginning to clean up their coal-fired emissions and so any respite will be brief. You should not conclude that burning dirty coal is a solution to global warming. The atmosphere sheds sulphur quickly but the carbon stays for centuries.

Figure 11 traces trends only since 1960. In his 2007 publication, Glenn Beck disputes that rising carbon emissions have steadily increased atmospheric CO_2 since the onset of the industrial revolution. He claims that even recent trends were highly variable: in 1935, values for ppm were 320. They suddenly went up to 430 ppm in 1937, stayed around that plateau from 1937 to 1948, and suddenly fell back to 320 ppm by 1950. If his estimate of 430 ppm for 1937 were correct, that figure exceeds the 400 ppm of today, and the link between rising carbon emissions and rising atmospheric CO_2 would be severed for good.

Georg Hoffman published a critique of Beck that same year, finding fault on two levels. First, Hoffman doubts that the biochemical changes implied by the scenario are even possible. Beck posits that between 1935 and 1937, the atmosphere gained an additional 110 ppm of CO_2 (430 – 320 = 110). That equals an added 230 gigatons of carbon (110 × 2.08 = 228.8; for the rationale behind the 2.08 multiplier, look back to Box 1 on page 27). Hoffman asks how such a thing could happen? In 1936, did the oceans and vegetation, which normally absorb almost half that amount of carbon from the atmosphere per year, simply shut down for 2 years? If not, certainly nothing new was shooting that amount of additional carbon into the atmosphere. Then, presumably the oceans and vegetation suddenly recovered their absorption capacity for a dozen years. Then, beginning in 1949, they outdid themselves: their usual capacity to absorb 112.5 gigatons annually doubled for 2 years. All of this is incredible, of course.

Second, Hoffman points out that Beck's recorded measurements at early years (1825 and 1857) were made at a time when scientists were unaware of factors that could bias measurement of CO_2 in the atmosphere. You have to allow for seasonal and diurnal cycles, the difference between terrestrial and marine sources, the difference

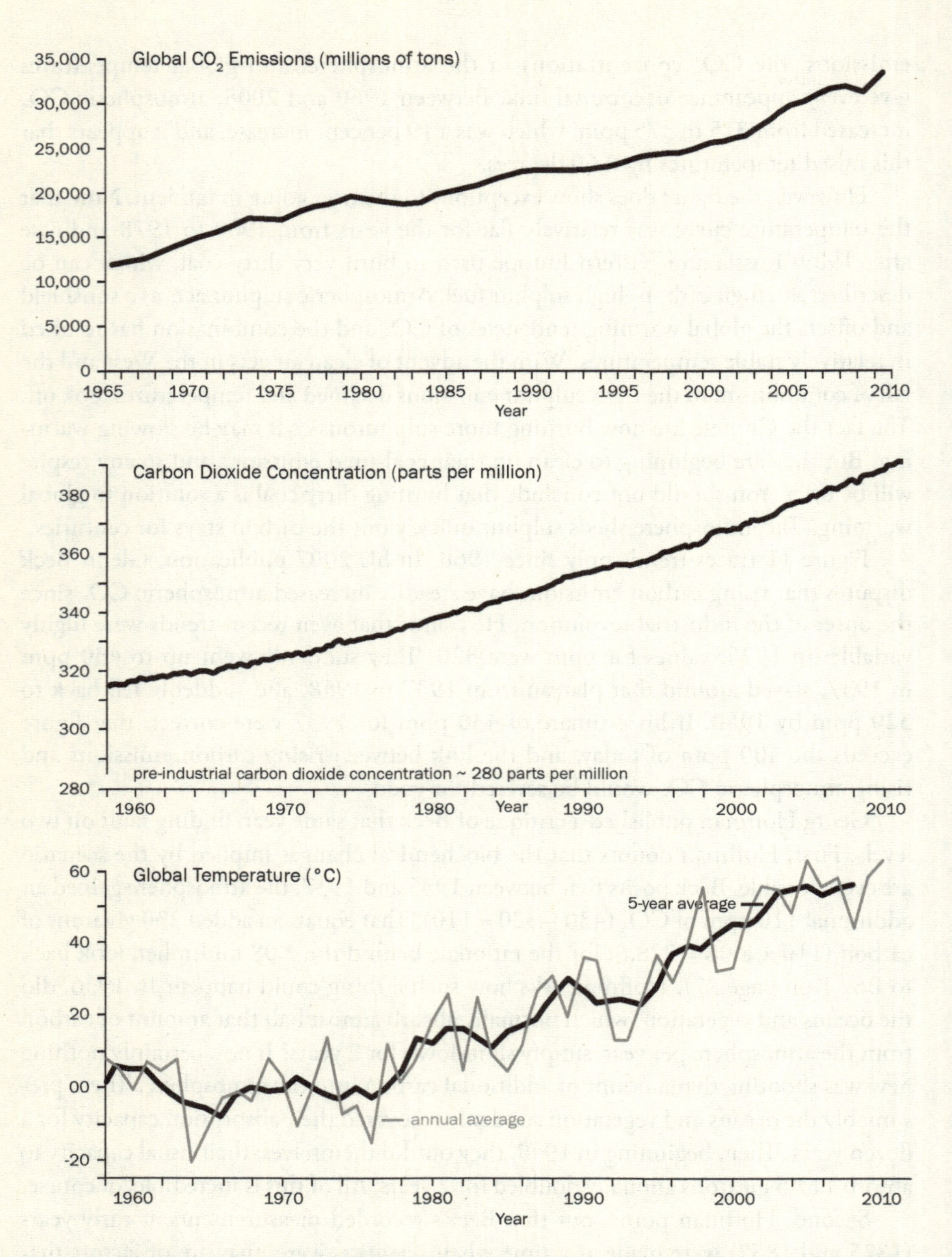

Figure 11. Recent trends in carbon dioxide emissions, carbon dioxide concentrations in the atmosphere, and global temperatures. http://earthobservatory.nasa.gov/Features/CarbonCycle/page5.php. THANKS TO NASA Goddard Space Flight Center.

between forests and plains, and elevated pollution in cities and their vicinities. Many of the earlier samples were collected in cities like Paris, Copenhagen, and Dieppe, or in areas surrounded by heavy vegetation. Collecting air samples is a fine art and pictures of 19th-century apparatuses show flawed samples being taken.

A Final Question

My students often ask me a question that troubles them: 'Even if atmospheric CO_2 is rising, perhaps it has nothing to do with us; the carbon emissions humanity adds to the atmosphere are small compared to the total that nature contributes each and every year.' What they say is true in the sense that nature does release a huge amount of carbon into the atmosphere each year. Why does the puny contribution of human beings make so much difference?

Table 1 is a balance sheet of what happened to the world's carbon during 1999. These are the only data I could find. Since nature's contribution to atmospheric carbon is stable, although the data are 17 years old, they are illuminating:

Factor	Carbon flux into the atmosphere (gigatons per year)	Movement of carbon out of the atmosphere (gigatons per year)
Soil organic matter oxidation/erosion	61–62	
Respiration from organisms in biosphere	50	
Deforestation (humans)	2	
Fossil fuel burning (humans)	4–5	
Incorporation into biosphere through photosynthesis		(110)
Diffusion into oceans		(2.5)
Net	117–119	(112.5)
Overall Annual Net Increase in Atmospheric Carbon	+4.5–6.5	

Table 1. Comparison of the annual increase in atmospheric carbon and carbon released into the atmosphere by burning fossil fuels and deforestation in 1999. http://soilcarboncenter.k-state.edu/carbcycle.html. THANKS TO Soil Carbon Center, Kansas State University.

1. The annual human contribution to atmospheric carbon was tiny. Deforestation plus burning fossil fuels contributed 6.0–7.0 gigatons out of a total of 117–119 gigatons (only 5.5 per cent).
2. But note how closely the human contribution (6.0–7.0 gigatons) matches the total increase in atmospheric carbon (4.5–6.5) plus the diffusion of carbon into the oceans (2.5). It falls about 1.5 gigatons short but we, after all, are responsible for some of the contribution of soil erosion.
3. Even though we have been concentrating on how atmospheric carbon increases temperature, I do not want you to forget the oceans and the effect that more and more atmospheric carbon may have on them.

The reason the human contribution makes so much difference is that what nature adds and subtracts each year balances out; at least when you count in what the oceans absorb.

Farewell to the Sceptic

I have tried to be fair. All sceptics force scientists to defend their hypotheses and climate sceptics have done just that. They have made us explain why the polar glaciers did not begin to diminish before 2008, reconcile our models with physics by evidencing the increased presence of water vapour, specify why the last 800,000 years offer unique evidence of the connection between atmospheric CO_2 and Earth's temperature, and refine our methods of measuring temperatures over the last 1,200 years. It is arguable that their final objections have only been answered over the last few years. Nonetheless, between 1999 (when the hockey stick reconstruction appeared) and today, their position has lost credibility, step-by-step.

CHAPTER 5

PREDICTIONS

When I visited Earth Sciences at Oxford in 2013, the researchers there told me that their models, assuming the persistence of current trends, predicted that temperature increase would be 6 degrees by 2100. As for the glaciers, the Greenland and West Antarctic glaciers would be gone, and there would be sea-level rises of about 12 metres. The East Antarctic Glacier would still exist. However, they added that, even if we capped atmospheric CO_2 at 1000 ppm in 2100, temperatures would continue to rise and all three polar glaciers would be gone by about 2300. The Oxford model takes into account all sorts of subtleties. For example, the temperature rise would not affect Earth evenly. Over the remainder of this century, ocean currents could favour an area like Europe and retard temperature increases there by a degree or so. But such local phenomena would count for little against the global trend.

Oxford and the Historical Record

When we surveyed the last 800,000 years, we saw that a 40 per cent increase in atmospheric CO_2 caused a 7-degree temperature rise. Atmospheric CO_2 stands today at about 400 ppm and increasing it to 1000 ppm by 2100 would be a 150 per cent increase. The ratio of future-to-past increases ($150 \div 40 = 3.75$) would seem to indicate a temperature increase of over 20 degrees by 2100. No model predicts that.

The chief reason is the persistence of polar glaciers. Purists would be appalled at the following analogy, but imagine taking a chilled drink out of the fridge. You take it into a warmer environment, namely, room temperature. If it has ice cubes in it, the drink will warm far more slowly than if it does not. We have three huge ice cubes: the Greenland Glacier, the West Antarctic Glacier, and the East Antarctic Glacier. It is impossible for all of our polar glaciers to melt in the next 84 years. As long as glaciers exist, their very melt is a cooling influence as it runs into the oceans and they themselves reflect heat away from the planet's surface. We are force-feeding the system, pushing atmospheric CO_2 up to 1000 ppm in less than a century (a speed that nature could never hope to imitate). Therefore thanks to the persistence of the glaciers over any short period, there will be a lag in global warming.

If you look back to Figure 11, you can see that during our own time between 1960 and 2008, human activity increased atmospheric CO_2 from 315 to 375 ppm, which was a 19 per cent increase, and this raised temperatures by 0.60 degrees. If we reach 1000 ppm by 2100, you get a 167 per cent increase from today. That rate would raise global temperatures by an additional 5.27 degrees by 2100 (see Box 2). We have already experienced a 1-degree rise from 1850, so you can see why the Oxford estimate looks about right. In fact, the relationship between ppm rise and temperature rise is not calculated in quite that mechanical way.

Box 2: Calculations

(1) A 19 per cent ppm increase = a rise of 0.60 degrees.
(2) Going from 375 ppm to 1000 ppm is a 167 per cent increase ($625 \div 375 = 1.67$)
(3) $167 \div 19 = 8.79$
(4) $8.79 \times 0.60 = 5.27$

Sea-level Rise

The potential contribution to sea-level rise if the three great polar glaciers disappeared is East Antarctic 58 metres, West Antarctic over 5 metres, and Greenland 7 metres. No one knows what the mass of the East Antarctic Glacier will be in 2100. I will assume that it will match its present size, which is to say that gains over the next few decades will be cancelled out by losses toward the end of the century. If this is so, and if the other glaciers disappear, that would raise sea levels by 2100 by at least 12 metres. I say 'at least' because warmth expands seawater (just as heat makes mercury or alcohol rise in a thermometer), and that would also push up sea levels. (Oddly, water also expands when it freezes because ice has large crystals. Otherwise ice would not float.)

The only projections of the effects of sea-level rise I could find were for the USA and ended at 8 metres (25 feet). Actually, I suspect 8 metres (rather than 12) may be about right. The disappearance of the West Antarctic and Greenland glaciers by 2100 is contingent to some degree on continued industrial growth, and its carbon emissions, until that date. However, as the erosion of glaciers begins to bite, later on in this century, it may well inhibit industrial growth. For example, moving cities inland would mean a fall-off in productivity. Ironically, the very erosion of polar glaciers may prolong their life a bit. In addition, if rising temperatures cause mass starvation in Africa and India, these regions will not support growth.

Such a rise would cost New Zealand its low-lying cities, namely, Napier, Hastings, and Blenheim, and much of Invercargill, the low-lying neighbourhoods of Christchurch, Timaru, and Dunedin (my home), perhaps one-half of Nelson and one-third of Auckland.

A rise of 8 metres would flood American cities as follows: Miami, New Orleans, Galveston, Norfolk, and Atlantic City (100 per cent flooded); Cambridge (Mass.), Savannah, and Charleston (80–87 per cent); St. Petersburg, Jersey City, Sacramento, Newark, and Tampa (50–70 per cent); Wilmington, New York City, and Mobile (36–41 per cent); Philadelphia, San Francisco, Portland (Oregon), Portland (Maine), Washington D.C., Providence, Seattle, and Tacoma (13–21 per cent).

My projection of an 8-metre rise by 2100 may seem alarmist. Some models predict a 3.3- to 5.6-degree rise in average temperature by the century's end, which makes the Oxford estimate of 6 degrees, on which the 8 metres is based, seem high. However, they assume lower levels of atmospheric CO_2 by 2100. Moreover, they were made before we began to experience the recent rate of polar ice loss. The shift toward loss from 2002 to 2006 was 300 gigatons per year; the shift between 2006 and 2014 was 500 per year. During 2014–15, just one year, it reached 3,000 gigatons, and during 2015–16 it rose to 4,000 gigatons. If that trend foreshadows the future, the cumulative loss will reach 4,800,000 gigatons by 2110. That is the total needed to eliminate both the Greenland and West Antarctic glaciers.

Ken Caldeira (Carnegie Institution Department of Global Ecology) brought a curious asymmetry to my attention. The past shows that when the world got colder, the sea fell by 20 to 40 metres per degree; but when it got warmer, the sea rose by about 14 metres per degree. Using simple arithmetic, if we warm the world by 6 degrees by 2100, the sea would rise by 84 metres. But remember, we cannot fast track the process in that mechanical way. We are hitting the world with atmospheric CO_2 at an unprecedented rate. Perhaps sea levels will respond more slowly to temperature rises over a very short time (250 years) than they did to the same rises occurring over a much longer time (1200 years). For example, in the short term the East Antarctic Glacier might still exist, while in the long term it would be gone. In other words, predictions have no exact precedent including my own 'radical' prediction.

Even if we assume a sea-level rise as low as 1.5 metres by 2100, the effects would not be negligible. The fate of American cities would run: New Orleans 88 per cent flooded; Galveston 68 per cent; Atlantic City 62 per cent; St. Petersburg 32 per cent; Cambridge (Mass.) 26 per cent; Jersey City 20 per cent; Charleston 19 per cent; Tampa 18 per cent; Wilmington, Tacoma, Savannah, Norfolk, New York City, and San Francisco 6–11 per cent; Sacramento, Mobile, Seattle, Portland (Maine), and Portland (Oregon) 3–4 per cent; Washington D.C., Newark, Philadelphia, and Providence 1–2 per cent.

Do remember that we are talking about land ice and not sea ice. Anyone who focuses on the latter at the expense of the former has an axe to grind. What they say is simply a distraction from the task of estimating the effects of global warming.

Melting Permafrost

When we looked at Earth's total emissions of carbon in a given year, we referred to carbon *in the soil* oxidising and boosting the carbon content of the atmosphere. Permafrost is defined as subsoil that remains frozen for at least 2 consecutive years. We are concerned mainly with long-frozen permafrost (for example, peat bogs). It stores huge amounts of carbon, that is, vegetation and animal remains that do not decay because they are frozen, rather like keeping something in the freezer. If temperatures increase over time, the permafrost thaws and microbes can get at this organic material and it decomposes, thus releasing its stored carbon to join our fossil fuel emissions and enter the atmosphere and oceans.

Permafrost with high carbon content is concentrated in an area that extends from Northern Alaska through Canada and Greenland and Northern Russia through all of Siberia. It is estimated to contain 1,672 gigatons of carbon. If all of this were converted into atmospheric CO_2, it would increase the atmosphere's CO_2 content by 804 ppm (see Box 1 on page 27 for the conversion equation). I should warn that this amount is not to be added to my (or any other) projection for 2100. Models have already factored in the contribution of melting permafrost. I am emphasising it simply to show that as temperatures rise, the fossil fuels we burn gain important allies that help to boost atmospheric CO_2.

How quickly the contribution of permafrost would grow depends on how quickly temperatures rise. Edward Schuur and his colleagues stated back in 2008 that permafrost would begin to melt gradually and amount to only 62 gigatons of carbon (4 per cent of that stored) being added by 2100. Their scenario is conservative as to when grave consequences (say sea-level rise) would occur. They believe that about 74 per cent of the stored carbon would be released sometime between 2300 and 2400. A few years later (2011), Kevin Schaefer's team, was less conservative, suggesting that perhaps 29–59 per cent of the stored carbon might be released by 2200.

Schaefer also speaks of a 'tipping point' at less than 20 years from 'now' (which makes it about 2030), after which the melt of all the permafrost will become inevitable. Others call this the 'point of no return'. By that they mean that once the temperature reaches a certain point, those higher temperatures, plus melting glaciers (less sunlight reflected), plus carbon from the melting permafrost will produce further warming, no matter how much we cut the level of our emissions. In other words, at a certain date, global warming may become a self-sustaining process. We will pay more attention to this 'point of no return' later on. There is disagreement as to just what year it may occur. In order to maximise consensus, I will suggest about 2050.

Acidification of the Oceans

The atmosphere and oceans exchange carbon. Where the air and water meet, CO_2 in the air dissolves into the ocean and forms carbonic acid. The oceans also release CO_2. Organisms all the way from microscopic plants (phytoplankton) to large marine mammals recycle CO_2 into the atmosphere when they breathe. Some phytoplankton and other organisms sink into deep waters when they die and marine bacteria convert them back into CO_2. Ocean currents drag this deep water to the surface and it vents CO_2 to the atmosphere like smoke escaping through a chimney.

Before the industrial revolution, the oceans vented CO_2 to the atmosphere in balance with the carbon the oceans received. Today we are increasing CO_2 concentrations in the atmosphere so quickly that the oceans take more carbon from the atmosphere than they release, and carbonic acid is on the increase. As the oceans become more acidic, calcium carbonate is less available and it is feared that this will affect the food chain. Many organisms use calcium carbonate in constructing their shells and skeletons, not only small ones (eaten by larger fish such as salmon) but also mussels and oysters. There is a passionate debate about whether acidification is already at work depleting oysters in the north Pacific.

We saw in Chapter 1 that, in 2012, Baerbel Honisch of the Lamont-Doherty Earth Observatory presented the findings of 21 researchers. A 5,000-year hot spell 56 million years ago is the closest parallel to current conditions. At that time, greatly increased volcanic activity had pumped CO_2 into the atmosphere at an unusual rate, doubling the atmospheric concentration of CO_2 and the oceans responded by becoming more acidic. Looking at mud under the ocean floor, the team found that many corals and many single-celled organisms became extinct, which is indirect evidence that other plants and animals higher on the food chain died out as well.

The projection for the year 2100 is a rise in ocean acidity equal to the rise that occurred over that 5,000 years. If marine organisms had found it difficult to adapt when they had 5,000 years to adjust, they may find it rather more difficult within a span of 250 years. In Sam Dupont's 2008 experiments, where acid levels were raised by a comparable amount, less than one in 1,000 of the larvae of a temperate-zone brittlestar survived (compared to 290 in 1,000).

Back in 2005, James Orr and colleagues predicted that by 2050, vast areas of both the Southern and Arctic oceans will be so corrosive that seashells will dissolve. Another team, lead by Gretchen Hofmann in 2010, showed that acidification poses a host of challenges to sea life in every ecosystem, whether tropical, open ocean, coastal, deep-sea, or high-latitude. The coral reefs of the tropics and the arctic seas may be the first to suffer devastating effects. A word of caution was added in 2011 by John Pandolfi and his fellow researchers: corals may be more adaptable to rapid acidification than some believe. However, everyone agrees that drastic cuts in CO_2 emissions offer corals the best chance to survive.

Agriculture

Rising temperatures will have dramatic effects on what areas become more or less suitable for agriculture. Some will profit from rising temperatures but most people will not. This makes sense. In the past, most of the world's population was distributed on the basis of present sea levels and whether people could be fed by food grown locally. The modern world has modified this by food imports and exports, but nations that are undeveloped or have developed only recently are particularly vulnerable.

The British Met Office's Hadley Centre for Climate Prediction and Research has published the best report on winners and losers. It was released in December of 2011 and covers 24 nations. I will use the scenario called 'A1B'. It assumes that failure to cut emissions will raise the CO_2 content of the atmosphere to 740 ppm by 2100, the best fit for my prediction of something less than 1000 ppm. Some of that report's assumptions have been shown to be too modest – for example, it has the world's population peaking at 9 billion, rather than the 11 billion projected at present.

The 2011 report shows that Britain will do well, with 96 per cent of its land more suitable for agriculture. Germany will also benefit, with 71 per cent more suitable, as will Canada, at 61 per cent; Peru, at 60 per cent; and Russia, at 40 per cent. Northern nations like Scandinavia will see as much as a 30 per cent increase in wheat production and a 50 per cent increase in corn. Of all nations, Canada may benefit the most. Looking ahead to 2050, if it capitalises on its enhanced attractiveness for immigrants, its population increase should lead the developed world.

As for the percentage of land less suitable for agriculture, according to the 2011 report, the following will be the worst hit: Spain (99 per cent), Australia (97 per cent), Turkey (97 per cent), and South Africa (92 per cent). France will show a net loss of 51 per cent. Food production will decline dramatically in parts of Brazil, China, India, the USA, and Egypt. Less water (drought) may produce water wars. Virtually the whole population of Egypt will be hit by water shortages. Egypt is more militarised than the nations who plan to drain the Nile before water reaches it, namely, Ethiopia and Uganda. Similarly, Syria and Iraq deeply resent Turkey's plans to utilise more of the Euphrates-Tigris River system before it reaches them.

How nations might adjust to these developments will be determined partially by wealth. For example, the USA can import large amounts of food. It can also adjust its agriculture by diverting food away from exports and animal feed. Conversely, in India, there are already farmer suicides in areas where most of the groundwater has been exhausted. The farmers are totally reliant on the cash that comes with the harvest and if it falls short, they are without hope.

The New Zealand Ministry for the Environment has done its own study. Most of the country will find its agricultural land upgraded with faster growth of pasture. Warmer temperature will mean new crops veering toward the sub-tropical in Northland and Auckland. Less severe winters will force kiwifruit production to move south. Marlborough will get better wines and Otago better cherries and apricots. Sadly the West Coast will get even more rain than at present, as will the Chatham Islands. Despite the prediction of more rain on the South Island, the chances of more periods of drought for all of New Zealand are put from minimal to double. This may be ominous in that the predictions are based on the assumption of a 1-degree temperature rise (a total of 2 since the advent of the Industrial Revolution). What a total rise of 6 degrees would mean we do not know.

There is an additional threat to the world's food supply, as highlighted by a very recent (2014) study by Samuel Myers and his colleagues. They provide evidence that when food is grown in an atmosphere with CO_2 concentrations of 550 ppm, its nutritional value diminishes thanks to less zinc, iron, and protein. We are scheduled to reach 550 ppm at about 2054.

Habitable Land

Sea-level rise will provide no winners. By 2100, small island nations face extinction and, not surprisingly, Mohamed Nasheed, president of the Maldives, has been among the most vocal proponents of climate-change mitigation. Whether the Netherlands can be protected depends on just how high sea levels rise. In 2100, about 49 million more people will be at risk, mostly in Bangladesh, China, Egypt, and India (where 16 million people will be affected). There is also the flash flooding that occurs far inland. For example, major floods deep in India, close to the Himalayas, currently kill as many as 10,000 and the incidence of these major floods is likely to triple. Kenya will lead the way with an even larger flooding increase in proportion to its population.

By about 2350, assuming all three glaciers go and a sea-level rise of 70 metres, every coastal nation will lament how much it has shrunk. As Figure 12 shows, New Zealand would lose Hamilton, Gisborne, New Plymouth, Napier, Hastings, Palmerston North, Blenheim, Nelson, Greymouth, Christchurch, Timaru, and Invercargill. Auckland would be reduced to two small islands centred on Mt. Eden and One Tree Hill. Large chunks of Wellington and Dunedin would survive.

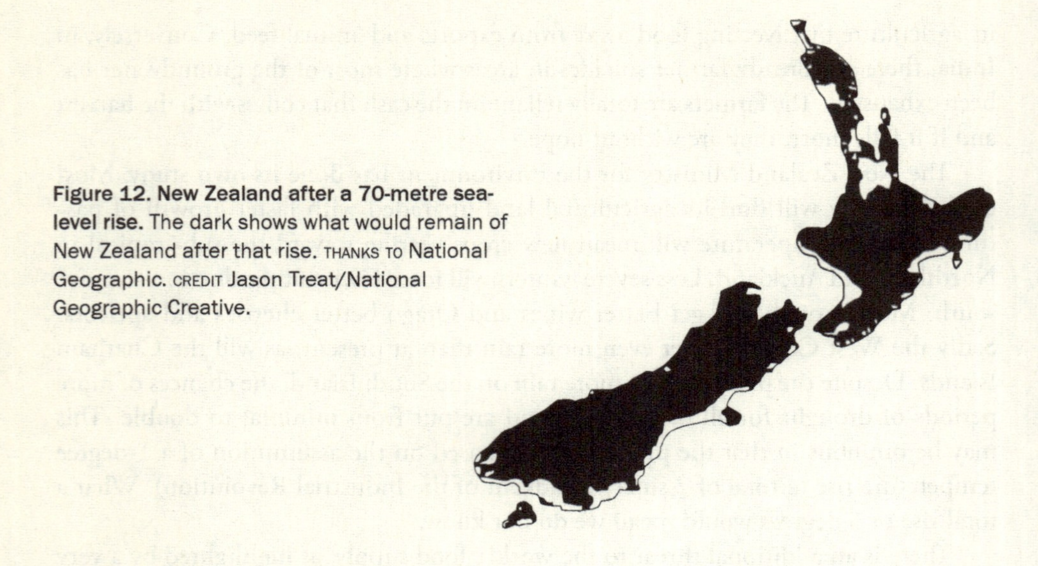

Figure 12. New Zealand after a 70-metre sea-level rise. The dark shows what would remain of New Zealand after that rise. THANKS TO National Geographic. CREDIT Jason Treat/National Geographic Creative.

Figure 13 shows that Britain would do no better. It would lose London, Liverpool, Carlisle, Norwich, Southampton, and perhaps Glasgow. The Scottish highlands would be largely intact and a strip running due south through the lowlands, the midlands, and hitting a south coast somewhat recessed. Most of Wales would survive connected to the main by a much narrower but substantial isthmus, while the Cornish peninsula would connect narrowly. Many of the eastern counties would be gone although a few fragments would become islands.

The USA would lose its entire East Coast (virtually up to the Appalachian Mountains and their plateau), all of Florida, the entire Gulf Coast (with the Mississippi River expanding so that all river towns south of Memphis would go under water), San Diego and San Francisco (much of Los Angeles could be salvaged if that was considered worthwhile), and coastal cities further north such as Portland, Seattle, and Tacoma. The northeastern provinces of China (inhabited by 600 million) would go, as would all of Bangladesh (160 million people). Germany would be a huge loser of its coastal areas. Some prediction maps can be deceptive unless you take into account the loss of coastal cites. For example, Australia looks intact until you realise that its coastal strip contains 80 per cent of its population.

Biodiversity

The work of Camilo Mora and his team has focused on the survival of plants and animals and in 2013, they introduced the concept of a climate departure date. Global warming means that at a certain date, the coldest year that occurs thereafter will be

Figure 13. Britain after a 70-metre sea-level rise. The dark shows what would remain of Britain after that rise. THANKS TO National Geographic. CREDIT Jason Treat/National Geographic Creative.

warmer than the hottest year before about 2000. For example, they put New York City's departure date at 2047: after that, every year will be warmer than the city's hottest year on record from 1860 to 2005. Mora believes we have already passed a point of no return in the sense that stabilising emissions starting tomorrow will only delay the departure date; for example, New York's date becomes 2067.

New York is typical of Earth as a whole, but Mora puts departure dates in the tropics as early as 2020. Unfortunately, the tropics hold most of the planet's species and these species are particularly vulnerable to climate change. He isolates 'biodiversity hotspots' defined as among the top 10 per cent of the most species-rich areas on Earth, and calculates the consequences of climate departure therein for 13 different groups, some marine (birds, cephalopods, corals, mammals, mangroves, marine fishes, reptiles, and sea grasses) and some terrestrial (amphibians, birds, mammals, reptiles, and plants). As he says, they will have to either move to a cooler climate, or adapt to the warmer climate, or become extinct. He believes this will apply to human beings. They, however, face the barrier of political boundaries: desperate people from Mexico will be unable to migrate north to escape the arid land they have come to live in. He does not discount the value of stabilising emissions: the longer species have to adjust, the more likely they can adapt.

New Zealand is not in the tropics but owing to its isolation is rich in unique species. The government makes no predictions about whether warming will diminish them, but it does assume that warmth will spread pest species (such as wildling pines).

Mora does not state the temperature rise that their models predict, but in an interview with Australian journalist James Nye, he says that present trends may mean a rise 'by as much as 7 degrees [Celsius]'. This is close to the 6 degrees predicted by the Oxford model as of 2100. In 2014, Ed Hawkins and a group of scientists argued that the departure dates have been stated with too much certainty and might well be later. Mora replied that his team's data and statistics are sound and that their conclusions stand. Amateurs like myself should regard the exact year of departure dates as tentative until others weigh in.

The Prospects for 2100

When we summarise these, we find that they are daunting. Sea-level rises, even at the minimum I have allowed, will dislocate many people. Some small island nations will cease to exist and the Dutch may be fighting a battle against the sea that they will ultimately lose. Wealthy nations like the USA will be able to adjust to the changes, and also compensate for reduced food production, but it will tax the country's resources. As we've seen, some other Northern Hemisphere nations may be relatively better off, Canada in particular. Britain may find that its increased food production counts for less than the negative effects of flooding. Japan's increased crop yield must be offset against reduced harvest from the sea. A fishing nation like Iceland may find its major export gone. Flooding, drought, and less food production will cripple India. Nations that will be badly affected by at least one of these include Bangladesh, Egypt and the Middle East, Spain, Australia, Turkey, and South Africa. Some places (mainly the far North and far South) will grow more food but the breadbaskets of today will suffer. We gave no fully adequate projections for New Zealand but clearly it will become like the tropical plus subtropical regions of today.

And there's one consequence only hinted at so far. Today's famines in East Africa are appalling but localised. If current trends continue, by 2100, the continent's total population will have risen from 1.1 billion to 4.4 billion people (making up 37 per cent of humanity). Mass starvation as a result of climate-induced famine promises to be both a consequence and a solution to the problem of more people than food. It will certainly multiply the number of those seeking a better life in Europe. During 2015 the total seeking asylum in the European Union was about 800,000. No one wants to think about what will happen when global warming multiplies this by 10 times or 20 times because everyone fears to say that machine guns are likely to be the answer: justified on the grounds that when a few thousand have been killed, the rest will stop coming.

The Point of No Return

These grim prospects, taken in isolation, actually paint too optimistic a picture. In 2100, people may find that their nation had long ago passed the 'point of no return'. I refer to the year at which a vicious circle begins to operate: another degree of temperature increase causes sufficient glacial melt such that reflected sunlight from the glaciers drops, which causes irreversible permafrost melt and the release of its stored carbon into the atmosphere, all of which cause more temperature rise. The dynamic interaction of these three means that even if we eliminate our CO_2 emissions at that point, the process will have spun out of control.

You can see that when climate scientists use the phrase 'point of no return', they are being literal. After a certain year, temperature rises will be inevitable, at least for a matter of centuries. Most scientists set 2050 as the year. That is when we seem destined to reach a level of CO_2 in the atmosphere at 500 parts per million. This is not to say that if atmospheric CO_2 goes even higher (from 500 ppm to 1,000 ppm by 2100), this extra will do no harm. If we can do anything to reduce the amount of carbon that gets transferred from the atmosphere to the oceans, the oceans will be far better for it.

The Ultimate Future of Humankind

Throughout the world, people will abandon city after city but they will build new ones. Patriots will lament losing much of their native land to the sea, but nation after nation has lost territory, sometimes in war (a more bitter pill), and kept its morale – witness the history of Germany and Austria and Russia. The USA will still be much bigger than the original 13 colonies and Russia larger than the Duchy of Muscovy. What is left of the population of Spain can take pride in the fact that its culture has been transplanted to Latin America.

Can we survive famine in Africa and the Middle East and India? Who can doubt that? Most of the rest of the world will still be eating a lot better than they did in the not-too-distant past, and those who eat have never died out because of sympathy with those who did not. For many the main change will be less meat and less seafood on the menu. Land wars over water resources will gradually be settled by the victories of the strong, hardly new in human history. If Americans needs Canadian water, they may discover that Canada has hidden weapons of mass destruction. The human race will go on living and loving and thinking new thoughts. There will be charity. During the Great Depression of the 1930s, an African tribe sent the USA the sum of $US 1.26 in foreign aid.

Some of those I talk to tell me I should be a detached observer of history. In Aldous Huxley's 1928 novel *Point counter point*, there is a conversation between Everard Webley, a fascist, and Lord Edward Tantamount, an eccentric peer who devoted his spare time to becoming one of the world's leading biologists. Lord Edward tells Webley that his political preoccupations are trivial: the real danger is that the human race is exhausting Earth's store of phosphates, coal, petroleum, and nitre. In exasperation, Webley asks whether he doesn't want what the fascists can offer: some alternative to a communist revolution. Lord Edward replies: 'Would a revolution reduce the population and check production? Then certainly I want a revolution.'

I want our future to be less fraught. At the end of this research, I felt that something as bad as World War II was looming over humanity, not as horrible in the short run, but worse in the long run. World War II also had some winners: many Americans lost a loved one but survivors enjoyed post-war prosperity; many neutral nations prospered; a colonial world discovered that the way was open to independence. What I want is nothing less than progress. The future does not have to be one that interrupts the march of humanity to a recognisably better world. As we will see, I believe that there are solutions. But I believe these will emerge only if we shed certain illusions. Those illusions are the substance of the next chapter.

CHAPTER 6

CAN WE ALTER THE FUTURE?

The greatest illusion is that the nations will agree to cut their carbon emissions in time to avoid the point of no return. In 2010, Roger Pielke published a brilliant book called *The climate fix.* He showed that we are improving the carbon efficiency of our economies. That is, for every $100 of goods and services the world produces (world GDP), we release less and less carbon into the atmosphere. Between 1990 and 2010, the rate of carbon released fell by an average of 1.3 per cent per year. However, world GDP rose by 3.45 per cent per year. It is clear who is going to win this race: unless the annual rate of de-carbonisation rises to 3.33 per cent, over two and one-half times its present rate; or unless growth slows to 1.32 per cent, which would be a cause of universal despair.

Between 1990 and 2010, even though the carbon intensity of the world's economy fell steadily, world GDP doubled. Therefore, CO_2 emissions rose from less than 22 gigatons to 33 gigatons. The best way to appreciate where we are headed is to project trends to 2100. This projection requires a third rate: how much do human emissions add to the CO_2 content of the atmosphere? We already know that from Chapter 4: since 1965, we have added 70 ppm of CO_2 to the atmosphere while increasing our emissions by 22 gigatons. This is a rate of 3.182 to one.

Predictions Assuming Continued Growth

Figure 14 shows that we will pass 500 ppm (perhaps the point of no return) in 2050 and pass 1000 ppm in 2100 (plus a few months). The data in the table below the figure may surprise you. If growth continues at 3.45 per cent a year, the world GDP of 2110 will be at least 30 times what it was in 2010. Lest you think this excessive, there are 57 nations (plus others for which we have no data) whose total income per capita could be multiplied by 30 times and whose average income would still fall short of that for nations like Norway, Switzerland, Australia, Denmark, Sweden,

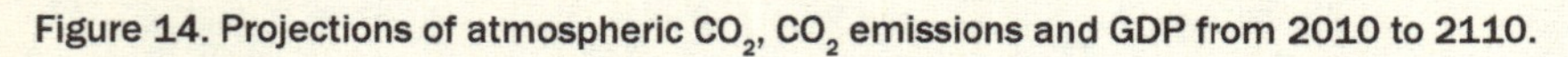
Figure 14. Projections of atmospheric CO_2, CO_2 emissions and GDP from 2010 to 2110.

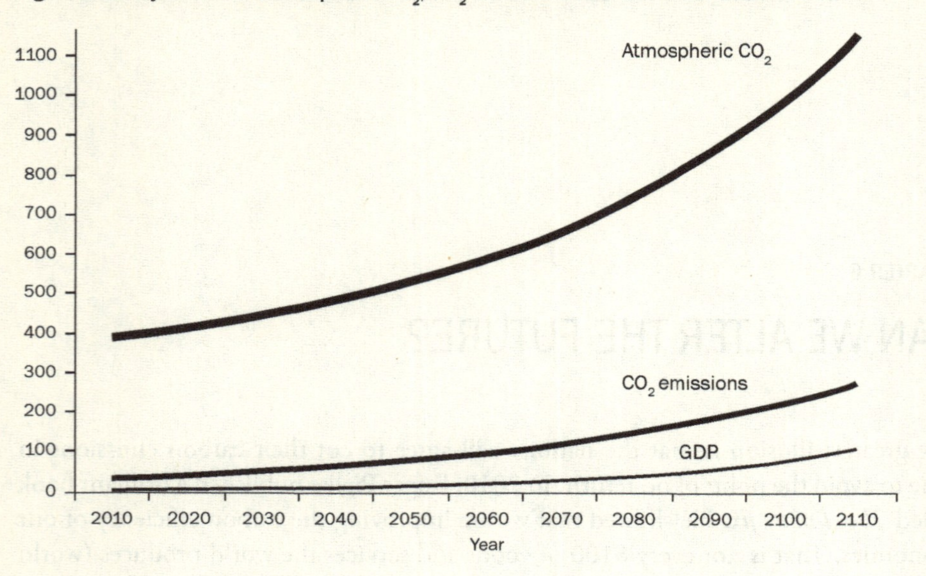

Year	CO_2 emissions (in billions of tons)	Atmospheric carbon (CO_2 ppm)	World GDP (in 10 trillions of 1990 dollars)	Efficiency (metric tons of carbon per $100 of GDP)
2010	33.02	390.00	3.58	9.22
2020	40.76	414.63	5.02	8.12
2030	50.34	445.11	7.05	7.14
2040	63.26	486.22	9.90	6.39
2050	76.76	529.18	13.88	5.53
2060	94.80	586.58	19.49	4.87
2070	117.08	657.48	27.36	4.29
2080	144.59	745.02	38.47	3.77
2090	178.57	853.14	53.99	3.32
2100	220.53	986.66	75.83	2.91
2110	272.36	1144.40	106.50	2.56

Canada, and New Zealand. Here are some nations and regions and the multipliers they would need:

- India 33 times
- Pakistan 40 times
- Bangladesh 60 times
- Southeast Asia 40 to 60 times
- Kyrgyz 50 times
- Tajikistan 58 times
- Haiti 66 times
- Afghanistan 88 times
- Most of Africa's nations would come nowhere close: 15 nations would need multipliers from 100 to over 200, and another 12 would need from 40 to 88 times.

Unless world economic growth persists over this century, it is hard to see how poverty can be alleviated. Equally disturbing, the final column of the table in Figure 14 shows that our scenario holds even if we maintain our present rate of improvement in terms of de-carbonising our economy. It assumes that we *continue* to cut carbon emissions by 1.3 per cent each year for every $100 of goods and services produced. We would face its consequences even if efficiency levels fell as low as 2.56 metric tons of carbon per $100 of GDP, huge progress on our approximately 9.00 tons today. In other words, we can maintain growth only if we get energy that is almost totally clean.

My projections in Figure 14 were calculated in 2010 and have been accurate thus far. For example, it predicted a rise of atmospheric CO_2 from 390 ppm in 2010 to 400 by 2015; as of 2 March 2016, it was 404.16 or just ahead of schedule. The long-range predictions do not take into account a factor we have discussed: that the consequences of global warming after 2050 may curtail our economic progress and thus lower carbon emissions into the atmosphere.

First, this is because I wanted to give projections that add concreteness to what the Oxford model predicts, which is based on the assumption that current trends are unaltered in any way. Second, I want to explore the possibility that deliberate actions by governments will cut the rate of carbon emissions dramatically and thus, avoid both global warming and a disastrous decline in the rate of increase of the world's GDP. Up to now, progress in the rate of carbon efficiency has been mainly due to individual awareness of the need to save energy (more solar panels and less wasteful cars) and market factors that have made low-carbon energy devices more economic.

But if the public has become aware of the need to curb carbon emissions, what about governments? I will argue that their response thus far has been feeble and that the recent Paris conference promises nothing better.

Pielke and the Past

At the Kyoto conference in 1997, the Kyoto Protocol was adopted. It marked the first international agreement to set out binding reduction targets for carbon emissions. Subsequent negotiations move from city to city: the 2014 sessions were at Bonn (Germany) and the 2015 sessions were in Paris. After each session, the IPCC (International Panel on Climate Change) issues reports, which are available online. Sometimes 'Kyoto' is used loosely to refer to the conferences about global warming and that convention is followed in this book.

As a result of the Kyoto conferences, nations set targets for cutting their greenhouse gas emissions. On the level of rhetoric, these targets are taken very seriously. On the level of doing something, the reality is very different. Back in 2010, Roger Pielke assessed what various nations would have to actually do to reach their targets.

In 2007, Australia set 2050 as the year by which it would cut greenhouse gas emissions to 40 per cent of their 2000 level. To do this, according to Pielke, it would have to replace all coal-powered energy with 57 nuclear power plants or some zero-carbon alternative. It is doubtful that public opinion would allow even one nuclear plant to be built. In 2008, the UK set 2050 as the date by which it would reduce emissions to 20 per cent of their 1990 level. It has the advantage that manufacturing is a shrinking portion of its economy. Even so, the UK would have had to replace coal and gas with 40 new nuclear power plants by 2015. All electricity generation would have to be de-carbonised by 2030.

In 2009, Japan set 2020 as the year by which it would reduce emissions to 75 per cent of their 1990 level. To achieve this would require: 15 new nuclear power plants far more efficient than existing plants; more thermal power; a 55 per cent increase in solar power; almost all new vehicles to be hybrid or electric cars; and all housing, old and new, with heat insulation. In 2009, 30 per cent of Japan's electricity came from nuclear power. As a result of the 2011 tsunami and the meltdown of the Fukushima nuclear power plant, Japan has debated abandoning nuclear power entirely. The effects of radioactivity and difficulty in disposing of spent nuclear fuel have alarmed the public.

All 50 of its atomic plants were shut down for safety inspections. By mid-2016, they had reopened four, one of which was immediately shut down because of leaks and another by a court order. The issue has become a political football. Former Prime Minister Junichiro Koizumi says that the country should immediately phase out nuclear power and that the current Prime Minister Shinzo Abe must take responsibility. For future reference, disasters aside, Koizumi stressed the problem of the disposal of spent nuclear fuel. This debate occurs within the context of Japan's stagnating economy. I suspect Japan will have little choice but to restart its atomic plants: its political elite does not feel they can compromise economic growth. The country needed $US 300 billion (5.26 per cent of Japan's total output) simply to repair the destruction caused by the tsunami.

Fukushima created anxieties that spread to Europe. On 30 May 2011, Germany announced a plan to shut all nuclear reactors by 2022. This decision to phase-out nuclear power has been called the swiftest change of political direction in modern times. At present, Germany relies heavily on coal power, about the dirtiest source available. Many environmentalists believe that coal will offset the phase-out of nuclear energy and that the result will be even more CO_2 pumped into the atmosphere. Industry is panicked by the prospect of higher generation costs for electricity. Despite this, Chancellor Angela Merkel believes that Germany can be a trailblazer for a new age of renewable energy sources. We will see.

In 2009, the US Congress set 2020 as the date by which it would reduce carbon emissions to 83 per cent of its 2005 level. According to Pielke, the USA would have to build 300 new nuclear power plants, or 200,000 new wind turbines, or replace all coal by natural gas, or find an acceptable mix.

India and China have many millions living in poverty. Rajendra Pachuari, head of the UN's International Panel on Climate Change, has been explicit: 'Obviously you are not going to ask a country that has 400 million people without a light bulb in their homes to do the same as a country that has splurged on energy.' China is more polite. It believes that its one-child policy made a major contribution to climate control. It claims that it prevented 400 million births that would have entailed almost 2 billion tons of carbon emissions per annum (some demographers put this at only 100 million births).

In sum, in order to get a 50 per cent reduction of the 1990 level of emissions, the world needed either 2 million solar thermal plants, or 8 million wind turbines, or 12,000 nuclear power stations (430 were operating at that time), or a mix of the three. For each nuclear power plant you drop, you need 227 solar plants, or 904 wind turbines. With targets as daunting as this, it is easy to see why little was done.

Paris and the Future

This brings us to October 2015, and the Paris conference. I believe the emission 'targets' set there will become increasingly irrelevant and mainly a source of acrimony. All we have at present is a statement of intent and even if all nations kept their word, temperatures would rise closer to 3 degrees than 2, and even 2 degrees gives a 50/50 chance of reaching the point of no return. The crunch will come when nations have to submit a detailed plan of how to cut emissions, and when they have to get these plans implemented by their own governments. The Paris agreements have had one salutary consequence. Independent scientists have updated Pielke's assessment by calculating what will actually have to be done from 2015 on.

For example, based on calculations from Joeri Rogelj (of the International Institute for Applied Systems Analysis), science journalist Fred Pearce asserts that an effective strategy is just barely possible. In a 2016 article in *New Scientist*, he argues: that between 2015 and 2050, we must limit the total number of gigatons of CO_2 put into the atmosphere to 800; and that between 2050 and 2100, we must achieve *negative* emissions by removing a total of 500. Temperature would rise to 2 degrees (the danger zone) by 2050, but then fall back to 1.5 degrees by 2100. The ultimate target for atmospheric CO_2 is set at 430 ppm. As for meeting this target, calculations accompanying Figure 14 on page 52 show that stalling emissions at a total of 800 CO_2 gigatons gives approximately 500 ppm of atmospheric CO_2 (2-degree temperature rise level); and then removing 500 gigatons gives 430 ppm (1.5-degree temperature rise level). In other words, Pearce's two steps probably would limit the ppm of CO_2 in the atmosphere to 430 ppm, the highest level acceptable.

I would amend Pearce's description of this scenario from 'barely possible' to 'not remotely possible'. His two steps assume virtual elimination of coal, oil, and natural gas well before 2050 and carbon-free generation of electricity. As African nations seek economic growth to try to feed their growing population, they will use coal as it is by far the cheapest source of energy, and they will be unwilling to pay the costs of importing energy. Even China plans to bring another 360 large coal plants on line by 2022, although it will relax its schedule after that date in an effort to control air pollution. In October 2015, it replaced its one-child with a two-child policy, so its population growth will be less inhibited.

Furthermore, there will have to be a huge breakthrough for transport: better batteries are needed to make electric cars universal. Since you cannot plug a plane into an outlet to charge it up, aircraft would have to be run on biofuels (which create their own problems in reducing land for agriculture). Air travel is so tough it was actually exempted from the Paris accords.

Also, by 2030, you would have to arrest deforestation, because if you do not it would undermine 80 per cent of the most optimistic plans for carbon capture (see Chapter 7 for more about this technology). This posits a huge tree-planting project that would cover an area as large as the continental USA. Any tree-planting will be swamped by present trends toward eliminating more and more trees. The rate of growth will send today's global population of 7.3 billion people up to 9.7 billion by 2055. This population increase will be concentrated in poor nations where people have to feed themselves with what is at hand, that is, they will have to slash and burn. This has already started, of course: between 2010 and 2015, world forests went down from 4,016 million hectares to 3,999 million, according to a recent report (2015) by Trevor Keenan and his colleagues. That loss is equivalent to three-quarters of the area of England.

Therefore, the hope for carbon capture of 500 gigatons post-2050 has turned from the land to the sea. Microalgae or seaweed farms could be located in the world's five great ocean gyres. Gyres are huge, relatively stable areas in each of the five major oceans that are created by global wind patterns and Earth's rotation. At present, they lack sufficient nutrients for seaweed growth and are called 'ocean deserts'. The plan is to seed the gyres with fragments of *Sargassum natans* (a species of seaweed) and use selective (no-nitrogen) fertilisers. This is a massive project. Enthusiasts often say that they would farm only 9 per cent of the world's oceans. But this amounts to 32 million square kilometres, or an area four times that of the continental USA – so these proposed farms are huge.

The seminal article about seaweed came from Antoine N'Yeurt and his team in 2012. He puts the cost of establishing the first open-ocean farm (in about 2028) at about $US 100 million and it would cover only 100 square kilometres. His ultimate target for 2035 is to exploit the full 32 million square kilometres possible. He gives no estimate of the scheme's total cost. However, unless there are economies of scale, at $1 million per square kilometre, the total investment would be $US 32 trillion). He grants that his goal is daunting from an occupied space and necessary technology perspective. I would add that there are additional indirect costs, such as the disruption of shipping lanes. Most of them go through gyres and avoiding gyres would lengthen travel times and multiply fuel costs.

To be fair, N'Yeurt and his team aim at reducing atmospheric CO_2 well below the 430 ppm considered safe and an algae farm the size of the continental USA (rather than four times the size) might be enough. This would be a minimal estimate in that his ppm target assumes not only CO_2 capture from the atmosphere by the seaweed, but also on harvesting the seaweed periodically and processing it to generate enough methane to replace all fossil fuels. Paris experts put the total cost at $US 270 trillion and at a quarter of the world's energy supply, in order for it to be fully effective by 2080. I believe that even the more modest version of this project is a non-starter.

Political Priorities

If the Republicans win the election in 2016, you can kiss American carbon targets goodbye. Even if a sane president is elected, any plan that includes phasing out fossil fuels will face the tremendous political power of the coal, oil, and gas lobbies and the workers they employ. In 2013, the coal industry employed 174,000 blue-collar workers, more than in 2000, plus the same number indirectly employed. In 2012, oil and natural gas employees numbered 570,000, an increase of 100,000 since 2007. If you triple these totals to take account of spouses and dependents, you get over 2.7 million people – the makings of a political nightmare for any American president.

No head of state wants to lose the next election by promising to cut growth and lower the nation's standard of living. This is not just a problem that affects democracies. No totalitarian ruler (or very few) can insulate his or her people from the modern world. Today millions of rural Chinese know that city dwellers have a better life than they do: what leader wants to tell them that progress must be deferred and ignite the powder keg that is the rural poor? As for the city dwellers themselves, they want to live more like Americans and their rulers know how much their legitimacy is in question. Read Wang Shuo's *Please don't call me human*, written in 2003. A government orator is portrayed as addressing a crowd but thanks to earplugs cannot hear what they are shouting: 'Go piss in a basin and look at your reflection.'

I have assumed that economic growth (and a better standard of living) is dependent on the continued use of fossil fuels. Many have argued that a shift to renewables will enhance growth at least in advanced nations. To be convinced, heads of state will have to: see growth faltering; believe that introducing clean energy will really restore growth; and believe that they can survive the political fall-out of the dislocation of millions of workers. All this must be demonstrable within 4 years (the normal term of office). Why rock the boat if the present drift is adequate? Once you have built a windmill, its maintenance requires few hands. This is a special case of a worrying trend: the technological innovations of the 19th century created jobs; recent innovations on balance eliminate them. And every promising scenario posits much more dependence on nuclear power, a proposal that faces fierce resistance in democracies. What possible chance is there that all of this is going to change?

Facing Reality

In a series of newspaper interviews, James Hansen, perhaps the most knowledgeable of the climate scientists, described the Paris conference as 'a fraud' and 'worthless words'. He is not alone in urging us to face reality. Not long ago (2013), Steven Davis, Long Cao, Ken Caldeira, and Martin Hoffert asserted two propositions that sum up this chapter.

First, even if current emissions were cut immediately by 20, 50, or 80 per cent, we would still pass the point of no return (atmospheric CO_2 at 500 ppm and an eventual temperature rise of 2 degrees by 2050). These researchers are dismayed by the long-term consequences. Their models agree with the science presented in Part A of this book. They show that maintaining a reasonable 'steady state' of Earth's temperature is controlled by a mechanism dependent on the polar glaciers (whether they exist), the chemistry of the ocean, and the composition of the atmosphere. Their models also show that altering these factors will mean new higher temperatures that persist for thousands of years. It will be as if a thermostat has been put on a higher setting and once that has occurred, it will take 12,000 years for temperature to fall back even 1 degree.

Second, there is no way of de-carbonising the world's economy that is viable within the next 50 years. The technologies that come closest to the near-zero emissions target all have limitations. Their report asserts: that CCS (carbon capture and storage) has not yet been commercially deployed at any centralised power plant, and that the existing nuclear industry, based on reactor designs more than a half-century old, is in a period of retrenchment, not expansion; and existing solar, wind, biomass, and energy storage systems are not yet mature enough to provide affordable base load power. (Thus, the unrealistic proposal for seaweed capture of CO_2 post-2050.) Indeed, they note that conversion of dirty technologies like coal to cleaner ones will initially raise emissions, as the infrastructure of the latter technologies is created and put into place.

Downsizing the Economy

Others bite the bullet and grant that de-carbonisation will reduce the productivity of advanced nations. They make an appeal to fairness: surely the fair solution would be to set a limit to world productivity that is compatible with cutting emissions and still affords a reasonable standard of living to all the world's peoples. For the sake of equity, they posit that developed nations would agree to diminish their productivity down to this level and that poor nations would be allowed to expand their productivity until they attained it. This assumes a degree of altruism that no developed nation possesses. It is not so much that they resent third world progress – some developed nations provide funds for economic development (although forfeiting advantageous terms of trade is not on the agenda). Rather it is a matter that welcoming the economic development of others is a far cry from planning negative growth for oneself.

To sugar the pill, there are economic models that show how the USA could have a lower level of productivity with full employment (for example, by adopting a 30-hour work week so that the same number of jobs would be shared out among more people) and a more equal distribution of income (through progressive taxation) that cuts the incomes of the rich and elevates the incomes of the poor. All my life, I have advocated that the USA should have a more equal distribution of income. This has not come to pass during a period of *rising* prosperity. It is hard to anticipate success when prosperity is *falling*. In passing, none of these scenarios show how we could get from the economy of today to the ideal economy of tomorrow without enormous dislocation, as the production of 'superfluous' goods and services Americans have been taught to think of as necessities is phased out. There are also problems of coercion: presumably holding two 30-hour per week jobs would have to be forbidden, just as bigamy is forbidden.

Even if the ideal economy existed everywhere, as the world's population rises, productivity must rise to cope. Once again, a limit to the world's population is a valid moral goal but the question is how it is to be achieved. While I was researching for this book, population projections rose from 9.5 to 11 billion. The lesson here is that you cannot control population among the poor by either education or coercion (unless you have an effective and totalitarian government like China). You control population by eliminating poverty and giving parents middle-class aspirations and the desire to use contraception (as the example of Europe shows). Only economic growth can limit population; limiting population to contain growth reverses cause and effect.

— PART C —

WHAT IS TO BE DONE?

CHAPTER 7

ALTERNATIVES TO CARBON-BASED ENERGY

Thus far, I have assumed that the main threat to economic growth at 3.45 per cent per year arises out of the post-2050 consequences of global warming. An alternative scenario is that sheer scarcity will cut productivity by then or soon after. This raises the question: How quickly are we depleting the world's resources? I also wish to explore in more detail the limitations of some recommended alternatives to carbon-based energy, particularly nuclear power.

Resource Depletion

An energy review published in 2011 by BP (British Petroleum) predicted that, based on consumption at the time, the world's supply of oil would last until 2056, natural gas until 2069, and coal until 2128. However, if global economic growth is to proceed, current demand must increase over the years. When that factor is built in (allowing for some progress in replacing fossil fuel-based energy with cleaner energy), the dates become 2049, 2051, and 2077, respectively. Actually, I believe that the date for oil could be postponed for 50 years, that is, until 2100 or beyond. The earlier date does not take into account new discoveries of oil (the size of Iraq's reserves were underestimated until recently) and far more important, shale oil and oil shale.

Shale oil is high-quality crude oil trapped in rocks through which it does not flow naturally. To access it, there is the extra expense of directional drilling and fracking – the use of hydraulic pressure (water pressure plus chemicals) to fracture the rocks. The latter is hardly environmental friendly. It poses problems such as contamination of ground water, chemical deposits on the surface, and some threat to air quality. There are huge deposits of shale oil in North Dakota (USA), indeed the twenty shale deposits in the USA contain far more oil than Saudi Arabia. Oil shale is different from shale oil. It is organic matter embedded in rock (any rock, not necessarily shale). You drill holes down to the oil-bearing rock (say 300 metres), use steel cables to heat the area up, and 3 years later, the organic matter has turned into oil that can be pulled to the surface.

There is some controversy about the world's conventional oil reserves but they may stand as high at 1.3 trillion barrels. The figure for shale oil has reached 1 trillion barrels and oil shale has been put at just above that, with 70 per cent contained in the Green River Formation covering portions of Colorado, Utah, and Wyoming. Recently, alarmed by shale oil production in the USA, Saudi Arabia has manipulated the price of oil downward to make extraction uneconomic. However, this has cut Saudi oil income drastically and the country cannot do this forever: the longer the low price persists, the greater the eventual correction.

In 2008, Pushker Kharecha and James Hansen urged that the oil in the ground not make its full contribution to global warming. They suggested that carbon taxes are needed to discourage conversion of the vast fossil resources into usable reserves, and to keep CO_2 beneath the 450 ppm ceiling. I suspect that carbon taxes may be used to conserve oil, but do not anticipate that governments will tax oil into non-viability, any more than they will do other things that trade growth for de-carbonisation.

If oil ever does become scarce, the main response is likely to be that we use more coal. If that is so, perhaps reserves of coal will run out before the estimated date of 2077. Trends over the last 2 years forced the BP energy review panel to lower its use-by date by 7 years. However, none of the estimates include the vast coal deposits in the Antarctic and the natural gas off its shores. If the price of oil rises, accessing and using these deposits will become economic and, as scarcity hits, the most likely thing to be renegotiated is the Antarctic Treaty that protects them. Ironically, as global warming thaws northern Alaska, it will make that region's large coal reserves accessible.

In sum, oil and coal reserves might be gone just before 2100. However, before the wheels fall off, high productivity should last well past 2050, the point of no return. After that the world economy may well implode. This would not prevent global warming but rather, would add its own contribution to our miseries. Put baldly, we need what seems impossible: a totally de-carbonised source of energy in the near future; and something that would arrest temperature rise now before atmospheric CO_2 has its full impact. In the next chapter, I will try to show that the impossible is actually possible.

As for other resources, there is no immediate problem for zinc, iron ore, and bauxite (to make aluminium). Although China is in a position to inflate the price of rare earth metals (for example, scandium, yttrium) due to its current near monopoly, deposits in Greenland will end that. However, the world's known phosphate deposits will be exhausted by the end of the century. This is a good thing, as phosphate fertilisers pollute water and, as we shall see, a better substitute is available.

The chief contender for future scarcity is water. It is needed both for agriculture and the production of pretty much every manufactured product on Earth. As for fossil fuels, almost 5.2 litres of water are needed to produce 1 litre of gasoline. The problem is not immediate stress on the planet's total volume of water, but the fact that water is unevenly distributed throughout the world. More than 2 billion people live in water-stressed areas (more will soon thanks to climate change). Actually, there are good prospects for extracting fresh water from seawater. A prototype 'water chip' was developed by Robbyn Anand and his colleagues in 2013, and it uses a small electrical field to remove salt. It looks energy efficient and may provide water on a massive scale. I suspect it will solve the problems of the rich rather than the poor. The latter may not be able to afford it and many poor nations have no seacoast from which to access saltwater.

Nuclear Power

Given the current limitations of alternative technologies (for example, nuclear power, natural gas, and carbon capture), it may seem extraneous to evaluate them. On the other hand, they have advocates who insist that these technologies are relevant. Some believe that nuclear power plants must be sold to the public and their leaders no matter how daunting the task. Nuclear plants split water (H_2O) into hydrogen and oxygen to produce carbon-free hydrogen. However, you need diesel fuel to run the equipment (to mine, split, and process the uranium ore) and additional energy to construct and decommission plants. The World Nuclear Association argues that the non-nuclear energy investment in a nuclear power plant is less than 1 per cent of the electricity it generates. Debate about how much of this dirty energy is required is ongoing.

The dangers of nuclear power plants are difficult to assess. It is claimed that nuclear power has caused far fewer accidental deaths per unit of energy than any of the other major forms of power generation. Perhaps, but certainty is not possible. The catastrophic disaster in 1986 at the Chernobyl Nuclear Power Plant spread contaminated material around the globe, but we cannot isolate its lethal consequences from other causes of death. Greater safety precautions, and their enforcement, might avoid such incidents in the future, but the earthquake-induced melt down at Fukushima shows that it is difficult to foresee every eventuality.

I have mentioned the problem of how to dispose of nuclear waste without prohibitive expense: spent fuel rods are dangerously radioactive for thousands of years. The best prospect is to store them deep underground in an area exposed to little groundwater and susceptible to little geological activity (otherwise, an earthquake might release the buried waste into the environment). The USA has selected a site at Yucca Mountain, Nevada. The Yucca Mountain is near a fault line and several volcanoes, but scientists testified that the fault seemed inactive and the volcanoes unlikely to erupt in the next 10,000 years. After massive opposition, Congress terminated the project on 14 April 2011.

Reprocessing has been discussed as a solution. In theory, 95 per cent of the remaining uranium and plutonium in spent nuclear fuel could be converted into a new mixed-oxide fuel. But this fuel can be 'burned' only in fast-breeder reactors, using the highly flammable liquid sodium as a coolant.

Frank von Hipple of the International Panel on Fissile Materials (IPFM) provided a good assessment of breeder reactors of this type in 2010. Scientists raised the possibility of such a reactor during World War II. By 1956, Admiral Hyman Rickover concluded that sodium-cooled reactors were expensive to build, complex to operate, susceptible to prolonged shutdown as a result of even minor malfunctions, and difficult and time-consuming to repair. And there is liquid sodium's highly flammable and explosive nature. Between 1980 and 1997, Russia's *BN-600* had 27 sodium leaks, 14 of which resulted in sodium fires. In 1995, Japan's prototype experienced a major sodium-air fire. Five French and British breeder reactors suffered significant sodium leaks, some of which resulted in serious fires. An important part of the problem is the difficulty of maintaining and repairing reactor hardware immersed in sodium. A large fraction of demonstration reactors have been inefficient.

France has the world's only commercial-sized breeder reactor. In December 1996, it ceased operations. Over 10 years, it had been shut down more than half of the time. The number of kilowatt-hours that it generated, compared to the number it could have generated had it operated continually at full capacity, was less than 7 per cent. France has huge stocks of nuclear waste it does not know how to store. It recycles only 28 per cent of its yearly output and three-quarters of this is done in Russia. The uranium in question is so contaminated it infects the plant that recycles it. It would seem that the Russians are willing to expose their workers to extra risks.

In sum, breeder reactors are costly to build and operate. Even optimists have set 2050 as a target for commercialisation. All nuclear reactors produce plutonium that can be used for nuclear weapons, but breeders are designed to separate out plutonium. The problem of denying access to weapon-makers would multiply.

Natural Gas

Natural gas is a formed when layers of buried plants, gases, and animals are exposed to intense heat and pressure over thousands of years. Sometimes it seeps from the ground: the Chinese used pipelines of bamboo to transport it beginning about 500 BC, mainly to boil water and extract salt. It is usually found in deep underground rock formations, often near coal or oil. It is most often used for heating, cooking, and electricity generation.

The main component of natural gas is methane. Although atmospheric methane becomes CO_2 within 10 years, the amount of CO_2 created is 44 per cent less than that created from the burning of coal. Methane is often described as the cleanest fossil fuel. Naturally, it is not nearly as clean as other energy sources like wind or solar power.

Therefore, it came as a shock in 2012 when Nathan Myhrvold and Ken Caldeira concluded that natural gas was almost worthless if we wished to slow down global warming during the 21st century. They posited that economic growth would persist, which is realistic, but assumed that the demand for electricity would stabilise, which is odd. They then compared natural gas versus coal as the sole means of generating electricity (electricity provides 39 per cent of human emissions). They took two important factors into account: that the total amount of atmospheric CO_2 persists over centuries; and that converting to a new technology will actually pump more CO_2 into the atmosphere during that period. The result: despite their assumption of no increased demand for electricity, temperatures would rise for at least 100 years; and only at the end of that time, would the emissions savings from natural gas kick in. Even then, switching to natural gas would cut the warming effect of electricity generation by only about 20 per cent.

Replacing coal with natural gas might have other long-term advantages, such as less carbonic acid in the oceans and therefore decreased ocean acidification. However, Caldeira is wary because any new investment money in the fossil fuel industry expands the size of that sector's political force.

Carbon Capture

Capturing and removing carbon from the atmosphere is beneficial because it lessens the total amount of CO_2 stored therein. Trees, like all plants, absorb CO_2 but planting is limited by the land, water, and nutrients available. As we have seen, forests are diminishing. Trees are capturing less and less CO_2, and we will do well if we arrest what has been a negative influence in carbon capture.

I have dismissed the creation of huge seaweed farms as impractical but there is an alternative, namely, enhancing the growth of the algae already in the oceans. The proposal is to deposit iron 'seeds' into the ocean that would act like a fertiliser and spur growth. This form of carbon capture would compensate for only 1 or 2 per cent of emissions because many marine organisms feed directly on the algae and ultimately return the CO_2 to the surface. Even more important, this strategy is a bad idea. We do not really want more algae depleting the oceans of oxygen and thus, creating dead zones where no animal life exists.

How algae do this is interesting. As algae grow, they consume all the oxygen in water and then die (suffocate themselves). Then naturally occurring bacteria start digesting the algal corpses but they have trouble doing so without oxygen. Therefore, they chemically bind up any scrap of oxygen that might still be left and use it to digest the remains of the corpses. That process ensures that the oxygen levels can't return to normal until the corpses sink to the bottom and are covered by other sediments. That can take weeks, and during that time oxygen levels are perpetually depleted.

Industrial factories and plants can be altered to use chemical means to capture carbon. The best are still experimental. These atmospheric-scrubbing plants, with specially designed equipment, use chilled ammonia to 'freeze' carbon dioxide-rich exhaust into crystals before it escapes into the atmosphere. The intention is for the captured CO_2 (except for a small portion that could be reused) to be piped to depleted oil and gas fields and un-minable coal seams, and stored underground. How much of it might leak out of storage is a subject of passionate controversy: the IPCC hopes for only 1 per cent leakage over a thousand years; Greenpeace notes that 1 per cent over a hundred years would return three-fifths of the stored carbon to the atmosphere.

If the leakage occurred all at once, it could be locally catastrophic if what happened in Cameroon in 1986 is anything to go by. There, a large leakage of naturally sequestered CO_2 rose from Lake Nyos and asphyxiated 1,700 people. A volcano had suddenly released the CO_2 from a deep valley hidden under the lake. Fortunately, no one would store carbon in such a location.

In a 2009 document, the UN International Energy Agency (IEA) endorsed new plants with carbon-capture capability and the modification of existing power plants (their lifespan is discouragingly long). The Agency gave carbon capture a 20 per cent role in its scenario to cut emissions in half by 2050. The IEA's programme would be expensive. It is hard to get an estimate of added costs: at one point the report put them at over $US 6 trillion (in 2010). The devil is in the detail. To build a new power plant with carbon-capture capability adds 50 to 100 per cent to plant construction and commissioning costs, and upgrading an existing plant is no less expensive. This percentage range refers only to the cost of the carbon capture equipment. In addition, more fuel is needed to compensate for the extra energy required to get the same power output. There is also the cost of the new pipelines and laying them, payouts to owners of the land under which they are laid, and the cost of the equipment needed to put the carbon underground.

The report acknowledges that the private sector will not undertake these costs. The governments of developed nations will have to defray them and negotiate arrangements under Kyoto to subsidise developing nations. The chances of this happening are about the same as any other arrangement of significant sacrifice under Kyoto – which is to say zero. The pace of change would have to be extraordinary. Over 40 years, 3,400 carbon-capture projects would need to be set up, that is 85 projects per year or one every 4 days. It would entail installation of 360,000 kilometres of pipeline, which is 10 per cent of the total gas pipeline existent in the USA. Whether there is sufficient safe storage for 3.3 billion tons of CO_2 on the planet depends on your views on the leakage problem.

Roger Pielke's 2010 estimations of the costs of various carbon capture strategies (see Chapter 6) included those for chemical means of atmospheric scrubbing. He hopes that this research will be pursued, as do I. But as he says, the current technology is largely speculative and certainly expensive and does not offer anything like a 'silver bullet'. It certainly offers nothing like the quick fix we need before 2050.

The world will continue to make progress on carbon capture, just as it will continue to make progress on wind turbines, solar power, electric cars, and painting roofs white. These developing technologies must become more efficient if we are to maintain the annual 1.3 per cent decline of carbon per $US 100 dollars of world economic growth already built into my scenario. They must take a huge leap forward in viability if our carbon efficiency rate is to match our growth rate. Even that would only hold our carbon emissions at present levels, and those levels will easily take us beyond the point of no return.

Politics and Ethics

We now know why politics and ethics forbid real progress. Our leaders believe that anyone who does what ought to be done will not lead for long (look in the mirror as to why). There is an overwhelming moral imperative looming over politics: do not kill growth and abandon the wretched of the planet. As for our tool kit for de-carbonising the economy, every year that passes demonstrates that the current kit cannot save us from passing the point of no return – that day when the feedback mechanisms of higher temperatures, melting permafrost, melting ice caps, and so forth, spin out of control and recreate the world without our permission.

CHAPTER 8

TWO SOLUTIONS

I do not believe in predestination. However, humanity seems to be sleepwalking down the road to clean energy. It leads away from carbon and toward hydrogen. Carbon becomes soot or CO_2 when combusted, while hydrogen becomes only water. Carbon has lost ground ever since 1800. Wood and hay were the major energy sources, and wood burns about ten carbon atoms for each hydrogen atom. In the 19th century, coal replaced wood. Coal burns one or two carbon atoms per hydrogen atom. In the 20th century, oil shifted the balance: kerosene or jet fuel burns one carbon to two hydrogen atoms. Natural gas is even better: its CH_4 has one carbon to four hydrogen atoms. Carbon provided 90 per cent of energy in 1800, but by 2100, given present trends, hydrogen will provide 90 per cent. That will not be enough thanks to the multiplication of the world's output of goods and services. We must try to get energy from hydrogen alone.

Moses and the Promised Land

Rob Ballagh, a physicist at the University of Otago, brought Ed Moses (what an appropriate name) to my attention. Moses heads the National Ignition Facility in California. In 2011, the NIF stated 'that its purpose was to create temperatures and pressures similar to those that exist only in the cores of stars.' It would do this by fabricating a 'tiny star', that is, an apparatus that would create hydrogen fusion in a tiny pellet. The process would be significant, of course, only if the fusion of the atomic nuclei generated more energy than the energy input that was needed to spark the reaction. This is called 'ignition'.

Figure 15 shows what Moses hopes to do. Press a button and 192 lasers release 500 trillion watts of energy, which is about 3,000 times the electricity consumption of the entire world. This energy strikes a pellet the size of a match head that contains a sphere of deuterium (heavy hydrogen) and tritium (even heavier hydrogen). The

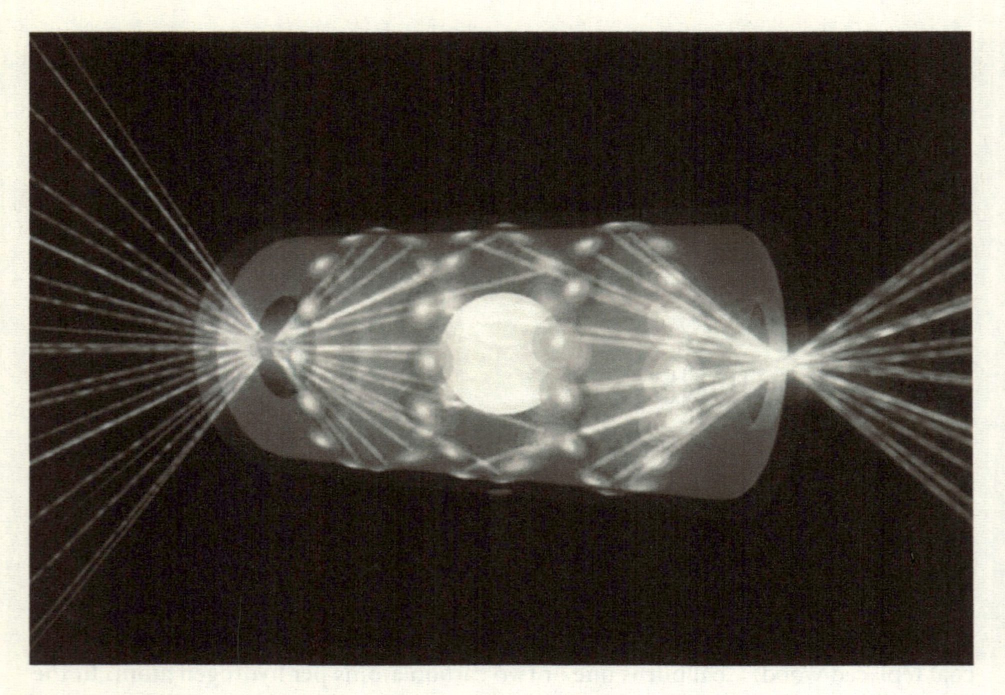

Figure 15. Beams at the National Ignition Foundation blast rays through a cylinder into a pellet of frozen hydrogen. https://lasers.llnl.gov/media/photo-gallery. THANKS TO Lawrence Livermore National Laboratory (National Ignition Foundation).

nucleus of a hydrogen atom has only a proton, that of deuterium has one neutron as well, while tritium adds a second neutron. The pellet is chilled to just a degree or so above absolute zero. The beams should compress the sphere so rapidly that it implodes, squeezing deuterium and tritium nuclei together until they fuse to form helium thereby duplicating the process that happens inside a star. Since the process is nuclear fusion rather than fission (which entails *splitting* the atom, as in a nuclear reactor), there is no troublesome radioactive by-product.

I used to wonder how the fusion of atomic nuclei produces energy. To say that we turn hydrogen into helium is not a sufficient explanation. Helium (plus an extra proton given off) has exactly the same number of protons and neutrons as the original hydrogen mix. And yet, mass must be 'lost' if you are to convert it into energy. Apparently, the original protons lose about 0.7 per cent of their mass in the process. That mass is what is converted into gamma rays (a lot of heat) and neutrinos (they don't count). If the entire pellet fuses, so that most of its potential energy is released, you will achieve the great goal of ignition; and you will have created energy of the sort that could drive a conventional turbine and produce electricity.

And yet, the facility's survival has been almost miraculous. Construction started in May 1997. The project had large cost overruns and in July 2005, US Congress actually voted to suspend construction. The NIF was saved by the argument that it would help the American military to simulate hydrogen bomb tests. The USA had suspended its nuclear testing in 1992 and since then, scientists have made computer simulations of how a weapon would explode. But from time to time, they want to test their simulations in the real world. The NIF is designed to create real nuclear explosions too small to count as nuclear tests within the meaning of the Comprehensive Test-Ban Treaty, but large enough to yield useful information. The search for clean energy is an add-on that costs extra.

By closing down the NIF, Congress had hoped to save $US 60 million, which is less than two-thousandths of one per cent of the Federal Budget. This is less than peanuts: Americans spend $800 million per year on peanut butter alone. However, Congressmen are dubious and some are climate-change sceptics and consider the money wasted. With the Congress looking over its shoulder, the NIF was under pressure to promise quick results. It missed a deadline set for September 2012, and it seemed certain that Congress would (retroactively) pass funding cuts for the year beginning 1 October 2013.

Then something wonderful happened. On 8 October 2013, the NIF reported that for the first time in human history, the amount of energy emitted by the pellet was greater than the amount that hit the pellet. This is still one step short of 'true ignition' because the energy that reached the pellet was only a small fraction of the total energy that lasers put into the system as a whole. Unfortunately, the pellet or capsule broke apart too soon for a large enough fraction of its hydrogen to fuse and for the process to become self-sustaining.

Moses is aware that one of the components of the pellet is highly unstable, namely, the tritium, but he believes it can be phased out in favour of the deuterium. He also asserts that the laser system for fusion energy will be compressed from its present size (huge) into 'a beam box' that could be transported by truck. If so, the pay-off would be enormous: the deuterium from 600 kilograms of water could provide energy equivalent to that yielded from 2 million tons of coal. We would have energy for a future equal to twice the lifespan of the sun. In Moses' view, the laser fusion process will be commercial by 2030 and carbon-neutral plants can begin to replace dirty-energy plants by 2050. Unfortunately, even if his predictions come to pass, existing coal- and gas-fired plants will not be eliminated for many years. Nonetheless, if every new plant and every replacement plant was carbon free, we might be putting a lot less carbon into the atmosphere by 2100 and very little after that.

The NIF cooperates with the UK's Atomic Weapons Establishment (AWE) and the Rutherford Appleton Laboratory. In 2005, the UK started a similar project called

High Power Laser Energy Research (HIPER), but it is not as far advanced. The NIF also collaborates with Spain's Instituto de Fusion Nuclear (IFN). Some scientists believe that improvements to NIF's equipment are necessary. China is studying its limitations with the possibility of building something better.

Princeton and Plasma

Princeton's Plasma Physics Laboratory runs the National Spherical Torus Experiment (NSTX). First operational in 1999, the NSTX was shut down in 2012 for an upgrade. This was completed in 2015. It uses another approach toward trying to compress deuterium and tritium and fuse them into helium. They exist as plasma, a gas heated so much that it conducts electricity, and the current generates a magnetic field around the plasma. This field creates an inwardly directed force compressing the plasma, which in turn strengthens the magnetic field, which in turn increases the plasma's density, and so forth, until you get a chain reaction resulting in the temperatures needed for fusion.

The NSTX consists of a huge box that powers a high-energy beam toward a circular container cradled by magnetic coils. It is lined with carbon tiles that can endure heat above 1 million degrees. However, the plasma within the container must eventually be heated to 150 million degrees, a temperature no solid material could withstand. Therefore, two magnetic fields (those generated by the current) separate the container and the plasma and confine the latter into the shape they call a 'spherical torus'. The plasma looks like a sphere with a hole that runs through its centre, which maximises the inward pressure on the plasma (see Figure 16).

Thus far Princeton has heated the plasma to over 60 million degrees. The upgrade should eventually double the amount of current flowing through the plasma. The team speaks of a commercial plasma reactor by 2050, 20 years behind the NIF's target for a commercial laser reactor. The recent success of laser fusion must not discourage research into plasma fusion. In 1997, Britain's Culham Lab generated 16 megawatts (MW) of power from a 24 MW input.

The Lab's head, Professor Steve Cowley, has high hopes for the International Thermonuclear Experimental Reactor (ITER) facility being built in the south of France. According to science journalist Ian Sample writing in 2014, France will have the first installation to produce more energy than it consumes. The ITIR is financed by the European Union, India, Japan, China, South Korea, and the USA. Its construction date is 2019 but fusion is scheduled for 2027, with a target of 500 MW. The next step would be building a commercial prototype, perhaps by 2030, the same year as the NIF target date.

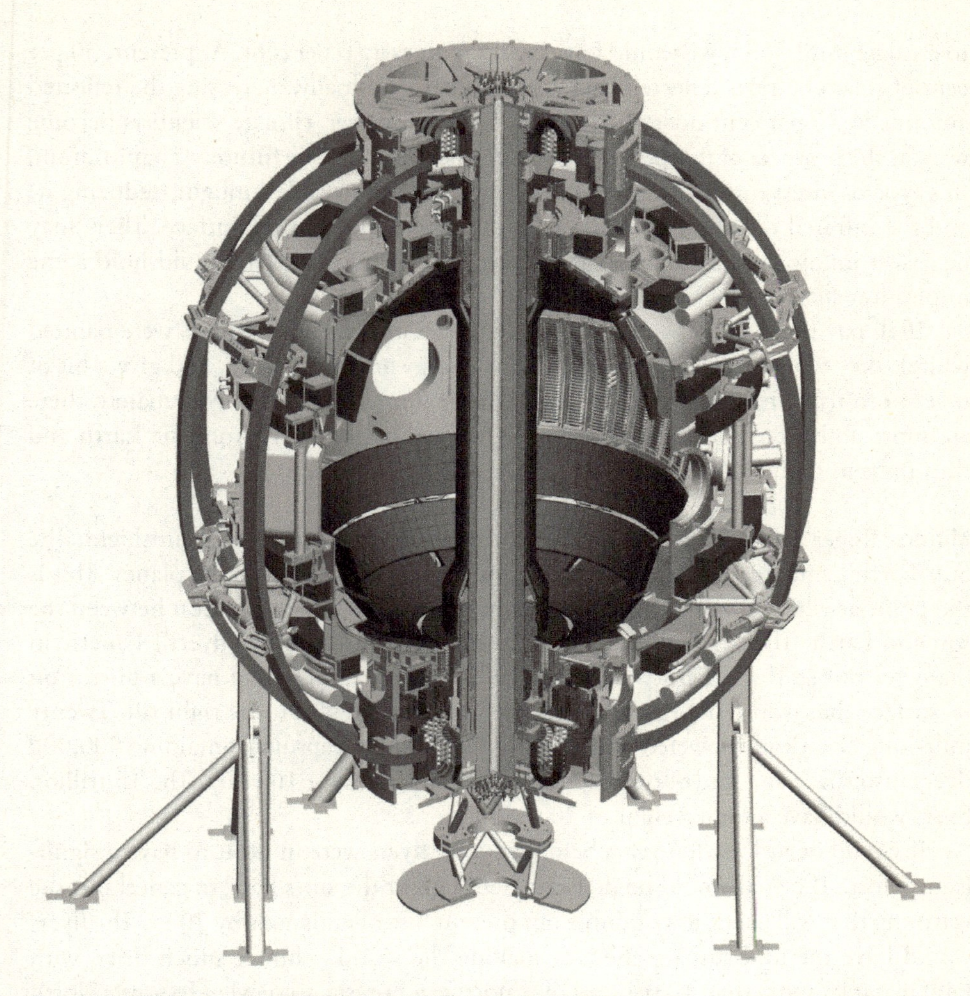

Figure 16. The National Spherical Torus Experiment (NSTX). https://www.computer.org/cms/Computer.org/dl/mags/cs/2015/03/figures/mcs20150300101.gif. THANKS TO The National Spherical Torus Experiment (NSTX) at the US Department of Energy's Princeton Plasma Physics Laboratory (PPPL).

Climate Engineering

Climate engineering refers to deliberate and large-scale intervention in Earth's climatic system with the aim of reducing global warming. If it could hold the planet's temperature at the present level, it would buy time so that clean energy can be effective even if it begins to dominate as late as 2100. In 2006, Roger Angel estimated that to counteract the effects of increased emissions from now to 2050, we would have to prevent an extra 2 per cent of the sun's light from reaching Earth. If emissions continue to escalate until 2085, we would have to block an extra 6 per cent. At present, 30 per cent of solar energy is reflected back into space, so gradually, increasing the reflected amount to 36 per cent does not look too risky. However, climate scientists become wary at the prospect of this percentage rising endlessly into the future. At a minimum, this type of intervention alters the balance between the visible sunlight (reducing it) and the infrared radiation (increasing it) that strikes the planet's surface. There may be as-yet unknown effects on weather patterns and rainfall that would hold some unpleasant surprises.

If all paving and roofs were made pale (for example, existing roofs were painted white), they would reflect more of the sun's energy and absorb less, and give a bit of relief from rising temperatures – but only in the immediate locale. Fortunately, there are more ambitious proposals that would reflect sunlight away from the Earth and thus prevent temperature rise.

Mirrors Roger Angel offers the most advanced proposal for a physical sunshield. The Sun-Earth Lagrangian point is about 1.5 million kilometres above the planet. This is the point at which small objects in orbit maintain a stationary position between the sun and Earth. The small objects he has in mind are discs (called 'flyers') 1 metre in diameter but so thin each would weigh only a gram. Each would have a mirror on its surface that would use the sun's radiation pressure to give it the right tilt. Twenty mile-long electrically powered artillery guns would shoot capsules containing 800,000 flyers into space at a rate of one capsule every 5 minutes for 10 years. The 16 trillion flyers would have a total weight of 20 million tons.

It would be at least ten years before enough flyers were in orbit to have a significant effect. The goal would be deflecting enough of the sun's light to cancel out the warming that will occur if we double our present rate of emissions by 2045. The flyers would have the unfortunate effect of making the world's climate much drier, with rainfall cut by more than 10 per cent over northern Eurasia, somewhat less over North and South America. The estimated cost is $US 6 trillion (circa 2012) with annual maintenance costs of some $US 100 billion. This may seem reasonable compared to what the world is spending on trying to cut emissions. But it would, of course, be additional to those costs if we ever hope to cut our emissions to zero.

Sulphuric aerosol Damon Matthews and Ken Caldeira have investigated a technique that would be both affordable (a few hundred billion dollars per year) and effective: pumping liquid sulphur dioxide into the stratosphere. In an interview with Eric Smalley in 2007, soon after releasing the findings, Caldeira said temperatures would go back to 1900 levels within 5 years. Differential spraying could restore the temperature patterns of the polar regions and tropics. According to one of Caldeira's recent

(2010) collaborators, George Ban-Weiss, the technique could not restore the rainfall patterns of 1900. Rain would sometimes shift from land to sea and some nations might experience more drought.

'Intellectual Ventures, the Stratospheric Shield' describes the mechanics. In the stratosphere, at an elevation of 30 kilometres, there is a layer of microscopic aerosol particles. Balloons would support a hose high in the sky pumping 100,000 metric tons of liquid sulphur dioxide (per year) into the stratosphere. Spray nozzles would atomise the fluid into a fine mist of aerosol particles, which would be spread by winds. This mist would be essentially clouds of sulphuric acid solution (sulphates) that would act as a sun shield. Since winds generally increase with altitude, the hose (or hoses) would be built to withstand a wind velocity of 60 metres per second. There are several ways to produce a continuous supply of sulphur dioxide: burning common sulphur-rich materials including wool, hair, rubber, and foam rubber; burning lignite (brown coal); and a slight modification of current jet fuel. The energy consumed by this strategy is trivial compared to the effect of its sun shield in lowering temperatures.

Sulphuric aerosols do not contain CFCs (chlorine, fluorine, and carbon) and therefore, pose no direct threat to the ozone layer. On the other hand, the sulphates would catalyse the conversion of existing chlorine near the Poles into an ozone-destroying form. This would do some harm. Research by Simone Tilmes and her colleagues published in 2009, indicates that this would delay the on-going repair of the ozone layer by some 30 to 70 years. Perhaps this is an acceptable price. Painters might find the clouds attractive: rather than a blue sky and vivid sunsets they would see a beautiful whitish blend of sky, cloud, and horizon.

Ships and sea spray John Latham (of the US National Centre for Atmospheric Research) made a proposal, which was operationalised by Stephen Salter and his team in 2009. A fleet of 1500 unmanned ships would drag turbines through the water to create sea spray. The sea spray would be pumped through micro-nozzles some 25 metres into the air as a sort of vaporised salt. At or above that height, turbulence would mix it into the clouds.

The spray would not make new clouds. Instead, it would cause existing clouds 1 kilometre above Earth to become brighter, which would reflect more sunlight back into space. There is a lot of fine-tuning to do. When a cloud's moisture is distributed over many small droplets, the combined surface area of the droplets is greater than if there were fewer bigger droplets. That greater surface area is what makes the cloud brighter and more reflective. The multiplication of the droplets depends on the formation of the nuclei of new droplets, which is the job of the salt. However, if too much seawater is sprayed into the cloud, the cloud becomes supersaturated, thus causing rainfall and the cloud becomes less reflective.

In addition, sea spray is not simply vaporised salt and water, but also contains other substances including viruses and bacteria. In 2013, a centre at the University of California (San Diego) received $US 20 million from the National Science Foundation to continue its laboratory experiments on health risks. Salter tells me that safety filters will eliminate all infectious agents. According to science reporter Deborah Brennan, the NSF will also study the total implications of this kind of climate engineering.

Salter's ships would sail back and forth perpendicular to the prevailing wind. They would be guided by a global positioning system, which would allow them to target suitable cloud fields, migrate with the seasons, and return to port for maintenance. Solar radiation would be reduced only over the oceans. Even before the NSF grant, in 2011, Govindasamy Bala and colleagues showed that if the spray were evenly distributed, India (where drought is a major worry), would be compensated for less rain by higher river flows. By selective spraying, it might be possible to increase rain in dry places and reduce flooding in wet areas. Endangered coral reefs could be targeted and the loss of ice near the poles discouraged.

Initial production costs would probably be $US 3 million per ship, or a total of $4.5 billion for 1500 ships. There would be a yearly maintenance cost of half a billion dollars for repair, replacement, and adding 50 ships per annum to keep up with increased carbon emissions. We should put a small fleet, perhaps 30 vessels, on the sea at once. This would be almost ridiculously cheap at less than $100 million. The US Navy's latest aircraft carrier cost $26.8 billion. Getting the ships on the water need not wait on the San Diego experiments concerning possible health risks. The first ships could spray in unpopulated areas and the effect on temperature monitored. Monitoring would be essential for determining the correct dose for the cloud formations and ironing out the inevitable problems posed by the mechanics of the equipment and the positioning system.

The best bet Let us assume that the disease factor is non-existent or minimal. At present, breaking waves send up the ocean's viruses and bacteria; and even if the filters fail, Salter's sea spray would add only 1 per cent to the present total. All in all, it appears that Salter's ships are by far the best bet. The mirror approach is far more expensive, the slowest to take effect, and unlikely to be accepted by nations that lose much of their rainfall. As for sulphuric aerosol, rain would wash the acid out of the sky every few years and, therefore, it would have to be continuously replaced; and the negative results for the ozone layer monitored. Although I am glad that Bill Gates is financing the project, it would be a harder sell than the ship-based sea spray option. People tend to reject departures from what they consider 'natural'. The spectacle of sulphur eliminating a blue sky might be unsettling, although a boon to millennial cults. For further reading, Eric Johnson's 2013 article contains a very good assessment of the three alternatives.

Other Good Things To Do

The member countries of the UN (essentially the great powers) should be the ones to fund long-term clean energy and immediate climate engineering. Needless to say, any nation that takes a step toward energy efficiency (that is, adopts substitutes for coal and oil) will help to sustain our steady progress toward shifting reliance away from carbon-based power. Such progress is essential to hold atmospheric CO_2 at as low a level as possible.

Solar and wind power Research is ongoing in these areas. In early 2014, an international team based at North Carolina State University and led by Maojie Zhang discovered a way to increase the efficiency of organic solar cells by more than 30 per cent, leading to lower costs and a much bigger market. Tim Flannery describes improvements in the cost and quality of windmills in his 2015 book *Atmosphere of hope*. Despite these advances, Bill Gates is sceptical: 'Well, there's no fortune to be made. Even if you have a new energy source that costs the same as today's and emits no CO_2, it will be uncertain compared with what's tried-and-true, and already operating at unbelievable scale and has gotten through all the regulatory problems, ... and how do you guarantee something is safe?' (from an interview with James Bennet in the November 2015 issue of *The Atlantic*).

Osmotic power Norway has investigated Pressure Retarded Osmosis (PRO) or osmotic power plants. In theory, these plants are feasible wherever freshwater flows into the sea, although they require a reasonably high concentration of salt in the seawater. The Norwegian modular plant consisted of 66 pressure pipes, a turbine, pressure exchangers, and a cleaning unit. The two different types of water were collected into different chambers separated by a membrane with very small holes. Enormous energy is created in the seawater chamber when seawater draws freshwater through the holes by osmosis. The plant would not discharge any pollutants into the atmosphere or affect the sea floor or any river habitat.

The Norwegian prototype began operating in 2009. However, plans for an operational plant were terminated in 2013. The problem was the membrane – it was not robust enough. Membranes are yet to be developed that are both thin and durable and experts put these at about 5 years from today (about 2021). Fortunately, Japan's Mega-ton Water System project has focused on producing better membranes. Its leading researcher predicts that Japan's rivers could be harnessed with this technology and produce enough power to replace five or six nuclear power plants. South Korea and Singapore are also funding research. The contribution to the world's energy needs would be small (about 1 per cent), but small nations with the right rivers could easily do much to end their reliance on coal. The hope is that one plant will produce almost 1,700 billion kilowatt-hours a year, or the equivalent of almost 700,000 tons of coal (that is 20 per cent of New Zealand's yearly consumption).

Rainforest preservation We are fighting a losing battle in saving trees but we should do as much as we can. The Amazonian rainforest locks up 10 per cent of the world's stored carbon. Its deforestation would significantly increase global warming. It is a self-sustaining ecosystem. If Brazilian farmers destroy 'a bit more' forest to bring land under cultivation, it may soon reach the point at which an irreversible process turns the whole area into a semi-arid savannah (Foley et al., 2007). The problem is that no one is willing to estimate what constitutes 'a bit more'. Things may be worse than they seem. Recent explorations show that some of what had been classified as rainforest is actually the result of past human intervention. It is not pristine and therefore, may not be integrated into the ecosystem.

The Brazilian portion of the Amazonian rainforest (about 65 per cent) lost 19 per cent of its area between 1970 and 2010. That country's government hopes to reduce the loss rate to 1.25 per cent over the current decade and to 1.2 per cent for each decade thereafter. It faces a powerful agricultural lobby that got protected areas reduced in 2012. The result was that the decline of deforestation was reversed in 2013 (back to the 2011 level). Other South American nations that contain parts of the rainforest are not going to be much help. The saddest case is Ecuador, which has announced that oil drilling will go ahead because of the failure of a UN-backed scheme. Its government was promised $US 3.6 billion (half the value of the oil reserves) as an incentive for forbearance. After 3 years, the total 'pledged' was $300 million of which only $13 million was actually deposited (that is, less than four-tenths of 1 per cent of the targeted sum). In Italy, I met a scientist from Ecuador who suggested that his nation was not unhappy that it could go ahead with development. Nonetheless, had the target been met, it could not have done so.

Brazil cannot be expected to curb the land-hunger of the poor and corporate interests, unless it can convince all of its people that they benefit from conservation. The UN should make Brazil an attractive offer: the government would receive a large sum (that would actually be paid) at the end of each year in which there was no deforestation. Now that rainforests and other vegetation can be monitored by satellite, it would be clear if the terms were met. If Brazil should ask for help with enforcement, so much the better.

The oceans and biochar For the time being, we cannot do much to cut the contribution atmospheric CO_2 makes to acidic oceans. However, we can protect the oceans from additional threats to sea life. Synthetic fertilisers are one of the great achievements of modern science, allowing us to feed the world's 7 billion people. Unfortunately, they pose a threat because they contain nitrogen and phosphorus.

In 1915, the Haber-Bosch Process was developed to convert inert nitrogen gas and hydrogen into ammonia, a reactive form of nitrogen beneficial to food plants. Jocelyn Kaiser, science journalist, summed up the problem nicely in 2001: 90 per cent of the reactive nitrogen leaks into the soil, water, and air. Livestock are an important link in the chain. Much of the world's grain goes to feed animals that produce mountains of nitrogen-rich manure. In the Netherlands, farmers must keep manure covered or plough it into fields. But even there, the nitrogen still escapes, carried off the fields by wind as ammonia or washed from the soil into streams. Fertilisers, either directly or indirectly, account for three-quarters of reactive nitrogen leakage. Nitric acid in the atmosphere causes acid rain and nitrous oxide supplements CO_2 as a greenhouse gas.

In 1927, the Odda Process was invented. Phosphate rock (with some 20 per cent phosphorus content) is dissolved in nitric acid and usually combined with potassium to form compound fertiliser, labelled 'N-P-K' according to the amounts of nitrogen, phosphorus, and potassium therein. Just as the combination of phosphates and reactive nitrogen encourages your lawn to grow, it encourages enormous growth of algae. As we have seen, when algae die, they create oxygen-depleted or 'eutrophic' zones fatal to most aquatic life. Over 32 per cent of the lakes in New Zealand and half the lakes in the USA are eutrophic. Back in 1975, A. J. Van Bennekom led a pioneering study that showed how the river Rhine affected Dutch coastal waters. The Gulf of Mexico has a dead zone (fed by the Mississippi) of over a million hectares (5,052 square miles). There is a larger one in the Baltic Sea. In 2014, the National Center for Coastal Ocean Science reported that the world's total had reached 550 dead zones.

Even the deep ocean is not safe. Some 20 million metric tons of phosphorus is mined every year and almost half finds its way into the oceans, thus adding to the acidification already created by their excessive carbon content. Work in 2003 by Itsuki Handoh and Timothy Lenton showed that at a time beginning 350 million years ago, mass extinctions of sea life occurred when critical thresholds of phosphorus flow into the oceans were crossed. Modelling suggests that a sustained increase of phosphorus flow is lethal if it exceeds 20 per cent of that produced by natural background weathering of adjacent coasts. They say that the present ocean is on the verge of dangerous oxygen depletion. More recently (2009), a team lead by Johan Rockstrom concluded that mass extinctions of sea life are likely over the next 1,000 years unless we set a limit of 11 million tons of phosphorus per year flowing into the ocean. The present figure is 9 million tons. Since the world's population will rise from 7 to 10 billion by 2050, the limit is clearly at risk.

The most promising alternative to nitrogen-phosphorus fertilisers is biochar. Biochar is a charcoal produced when wastes available on farms (wood, manure, crop residues) are heated in a closed container with little or no air at 'low' temperatures (below 700 degrees). After use as a fertiliser, it stays in the ground as carbon should. Each farmer or group of farmers can operate a low-tech kiln costing $US 5,500 that is transportable on a truck. None are available in New Zealand yet, but there are instructions for small-scale ones online. In 2010, Simon Shackley and Saran Sohi estimated the cost of a ton of biochar delivered to UK farmers commercially, which equalled $US 670. In 2012, the UK *Farmers Weekly* said that a ton of compound fertiliser would cost its subscribers $US 550. Therefore, initially, subsidies may be necessary. But the agricultural economies of affluent nations live on subsidies and if biochar meets expectations, it should be added to the list of subsidised commodities.

The benefits of biochar must be evidenced for a range of soil/crop types. Science journalist Chris Goodall produced a good summary in 2011 of Oxford (UK) University's project: The Big Biochar Experiment. Researchers were testing to see if the fertiliser is as effective in temperate climates as it is in the tropics, where it has sometimes increased yields enormously. A year later, Simon Manley reported a project that his company, Carbon Gold, had underway in West Africa. The cacao trees (they produce beans for cocoa) grown in that area are among the most highly sprayed crops in the world. When biochar replaced conventional spray as a top dressing, yields increased. Now the emphasis is on incorporating biochar into the soil at all stages (seedlings, semi-mature trees, and mature trees).

Kyoto Again

If post-Paris climate conferences stop making gestures and face reality, the situation is by no means hopeless. But we cannot drift for much longer. It is easy to blame the climate sceptics. But what about those of us who are alarmed? How many of us have given up on cutting carbon emissions as an adequate solution? How many of us are willing to follow James Hansen and face the necessity of climate engineering?

We need some sort of global planning. It would be wrong to allow clean energy and climate engineering to overshadow all else. We do not want to lose the Amazonian rainforest and we want to minimise the acidification of our oceans. But clean energy and climate engineering are fundamental to any effective long-term strategy. Laser fusion, and perhaps plasma fusion, offer a real hope of carbon-free energy before 2100. They are surprisingly cheap. Unless there are unforeseen negative consequences, Salter's ships can postpone temperature rise until that time. There is every prospect of stripping 2050 of its significance as the point of no return.

EPILOGUE

GOOD WILL TOWARD MEN

I was pleased to find that when I embarked on this enterprise, many of my friends, colleagues, and correspondents, both in New Zealand and overseas were interested in what I might conclude. I hope that this was because they thought that I wished humankind well and might approach the subject with an open mind.

No one can claim to be without preconceptions: I am on record as a Social Democrat. However, my motives for studying climate change had nothing to do with my ideology. I was simply curious about something that seemed terribly important. Although my friends were divided, I must confess that I suspected that climate change was significant. However, much that I discovered was an unwelcome surprise. I did not want to conclude that some form of climate engineering was necessary. As someone who had warned students against the excesses of materialism and had argued for a no-growth economy, a new society that was serious about the pursuit of truth, beauty, and social justice, I found it unpalatable to conclude that the best hope for the world's poor, and for limits on the size of the world's population, was sustained economic growth at the present level. Who would have thought that I would be searching for a scenario that promises a 30-times increase in the world's total productivity!

We must reshape the history of the remainder of the 21st century. Optimism is hardly in order, but it is perfectly possible that it could read something like this:

1. We get Salter's sea-spray ships on the water and lower current temperatures by say 1 degree – we will know we have enough ships when the polar glaciers stop melting.
2. If we get temperatures to that level, and do so no later than 2035 or 2040, we may avoid the point of no return – prevent the feedback mechanisms of higher temperatures, glacial melt, and permafrost melt from kicking in and overriding whatever we do.

3. If we make dramatic progress toward really clean power production between the years of 2065 and 2100, we can continue economic growth and yet hope to eventually reduce emissions to near zero.
4. In the meantime, emissions will continue to build up in the atmosphere and CO_2 will enter oceans and make them acidic – therefore, we must continue all the good things we are doing now that hold that increase to a minimum.
5. If economic growth has continued, Africa may become prosperous enough to join the rest of the world in its reproductive patterns, and the world's population really will level off at 11 billion.
6. The persistence of the CO_2 in the atmosphere means that the level will drop from its peak only gradually – but with zero emissions by 2100, we can monitor it and look forward to the day (perhaps after 100 years) when we can find something better to do with our now-redundant sea-spray ships.

It would be gratifying if everyone read this book, but it is rather likely that many people have things to do that put my book low on a priority list. If you have friends who will not take the time to confront the issue of climate change, tell them to keep an eye on three developments that are easy to follow and may arouse a sense of urgency:

1. Above all, follow the yearly data about the decline of the Greenland and West Antarctic glaciers (and do not be hypnotised by the size of sea ice). As we have seen, we can now monitor them in a way beyond dispute thanks to the GRACE satellites (the Gravity Recovery and Climate Experiment satellites the National Space Agency keeps aloft). The next 10 years will tell us a great deal. I predict that the yearly loss from polar glaciers will rise from 4,000 gigatons to over 5,000 by 2026. Unless people are camped beneath the glaciers and lighting fires, it is the best sign that our climate promises a fraught future. If I am correct, it is urgent to get Salter's fleet underway.
2. Ascertain whether world productivity falters or keeps increasing at a rate of 3 per cent per year. If that continues, only the most radical reduction of emissions will avoid the necessity of some climate engineering.
3. Read the summaries of the reports of the International Panel on Climate Change (IPCC), which are available online. They will update you on whether the industrial powerhouses of the world are setting, and abiding by, the kind of targets that would lower emissions.

Some have given up any hope of influencing the planet's climate. James Lovelock became a hero of environmentalists when he published *Gaia* in 1979. It argued that Earth was one self-regulating system in which human beings now play a key role, unique not merely because of their impact but because of their self-awareness. In his most recent book, *A rough ride to the future* (published 2014), and in subsequent interviews, Lovelock derides efforts to resist climate change and advises that people should retreat to climate-controlled cities and give up on large expanses of land that will become uninhabitable (offering little hope for developing nations here). He is ready to grant that humanity, like all organisms on Earth, has a limited lifespan: perhaps we can evolve ourselves from being organic creatures toward existing as computerised life forms.

When watching TV last year, I found it devastating to see one of the authors of the IPCC report offering hope that humanity could avoid the point of no return if only we progressively reduce our 2010 level of greenhouse gas emissions (the 2010 levels are much lower than the levels of today) by somewhere between 40 per cent to 70 per cent by 2050. It is this quixotic hope that has pushed so many advocates of a liveable Earth to despair. I do not think we should despair. However, the only hope lies in giving political elites the following: something they can do *without* committing mass political suicide.

The steps I propose do not exact a political price that is unbearable. Let me put them forward in the form of what every person can do to help. I will address this advice to my fellow New Zealanders but it serves as a model that anyone in the developed world can adapt. We should lobby our government to:

Go to the next climate conference with a coherent plan. It should ask the UN to put 30 of Salter's ships (generating ocean spray) on the water – the cost would be less than $US 100 million. The UN and all its agencies and funds spend about $30 billion each year, so this would be about three-tenths of 1 per cent. Russia has begun to demand that the Reports of the IPCC should include the 'insurance policy' of climate engineering! The New Zealand Energy and Resources Minister should approach the Russians to find out what they have in mind. Every mayor in every one of New Zealand's coastal cities should approach the Energy and Resources Minister with that message.

Our government should contribute a reasonable sum to a UN fund to compensate nations for not developing the Amazon. The total fund should be about $US 4 billion. If this were shared out between the USA, the EU, and other prosperous nations according to their gross national products, New Zealand's contribution would be $15 million. At present, its yearly overseas aid budget is $550 million.

Our government should study Japan's research on clean osmotic power plants. Even if better membranes are produced over the next 5 years, feasibility studies would have to be done for each of New Zealand's rivers. A river's average discharge rates (and its lowest flow rate) are crucial. How many of our rivers would qualify? Probably the Clutha, with a mean discharge of 533 cubic metres per second, would qualify and the Waikato, with a mean discharge 340 cubic metres per second, and several others as well.

As for individuals acting on their own, we could put on the web an organisation called SOS (Salter and Ocean Spray). It would not be limited to promoting Salter's ships. It would try to raise an international private fund to free Ed Moses and his collaborators from their yearly funding nightmare (it would be good if Warren Buffet or another philanthropist could take the lead). And it should advise people how to live as green lives as possible.

Ralph Chapman gives New Zealanders a good set of recommendations in his book *Time of useful consciousness: acting urgently on climate change* (2015): buy electric cars, demand electric buses, car-share, do more cycling and walking, demand development of wind and solar energy (and use solar heating), insulate homes, get rid of the energy-squandering Tiwai Point aluminium smelter. The energy savings of concerned individuals has always been greater than anything emerging from Kyoto.

Thinking people throughout the world should seek unity of opinion. In the 19th century, an international consensus on the wickedness of slavery made it indefensible. Sadly, politics stands in the way of a similar consensus about climate. We should try to put politics aside. A few on the right see global warming as a fabrication behind which lies a political agenda. They see a conspiracy to enhance the power of the United Nations, compromise national sovereignty, and facilitate regulation of the free market. The left sees fanatics so wedded to free-market ideology that they have become obscurantist, science deniers, and enemies of humankind.

I have tried to show climate-change sceptics that I acknowledge their contribution; and that I reject their case on honourable grounds. This book is not going to convince most of them but I hope I have injected some civility into the debate. Here is an anecdote from another unsettled debate, one about Russia's motives during the Cold War. A close friend once said: 'Assume you are correct about Stalin's ambitions being constrained by reality. But you concede you just could be wrong and if you are wrong, the consequences would be horrific. Surely it makes sense to arm for a worst case scenario and wait for Russia to show that it will settle for peace.' A powerful point! Surely no sceptic should discount the outside chance that we alarmed are correct (even though they think it highly improbable); and if we are correct, the consequences would be horrific.

In the absence of unity of opinion, can we not agree on the steps I propose because they satisfy the moral principle of 'do no harm'? Sceptics will say that money is being wasted. But what I propose is so cheap. Why not look upon it as an insurance premium? A very low premium that sceptics are nice enough to help pay, if only for the benevolent purpose of reassuring their more nervous neighbours.

As for hydrogen fusion energy research, who would not welcome a cheap and inexhaustible source of energy for its own sake? It would benefit free market economies as much as the Social Democratic economies of Scandinavia. If clean energy provides a foundation for continued growth, the alleviation of poverty, and a limit on population, everyone should be pleased. If stocks of coal and oil run out, or become very expensive to exploit, what harm would it do to be prepared rather than having to improvise? Even if you discount carbon as irrelevant to global warming, it is polluting the seas, so everyone should want an alternative.

This is not to say that psychological vetoes to progress come solely from the deniers. Many of the alarmed make a contribution. I have talked to them and they describe climate engineering as a 'band aid' that leaves the 'central problem' untouched. They describe it as counterproductive, as an expedient that nations would use to justify failure to curb emissions.

These people may be willing to face scientific reality but they cannot face another kind of reality. They are naïve about the politics and economics that doom what we are doing now to failure and worse, they have not assimilated that the best climate scientists (such as Hansen) have thrown in the towel. Let us say it again: even the best conceivable emissions cuts will not stop us short of the point of no return, unless they are accompanied by climate engineering to hold the line on temperature. I can put the entire message of this book in two sentences. If we are to multiply production by 10 or 20 or 30 times, just think, *really think*, about how spotlessly clean our energy will have to be. Now think about whether we are going to need to do something new to delay rising temperatures until that time.

Making people better informed is rarely enough to solve problems of great consequence, but I am convinced that the problem of climate change may be one of the rare exceptions. Serious people, media people, and the political elite must start talking sense to one another. Suddenly there may be a new point of no return generated by ideas. Physics, chemistry, politics, economics, and ethics will reinforce one another and a new feedback mechanism will be born. As it spins out of control, no argument can stand against it.

Every one of us wants to know the truth and, as so often, the truth is elusive. We should think of one another as human beings who need reassurance and good will, not stones and arrows. To those who calculate that they will die before the point of no return and care nothing for the next generation, I have nothing to say. They have declared war on humanity. They will ignore the wellbeing of the people of the future just as they ignore the misery of the people of the present.

The mind is not enough. Character must rise to the occasion. Many of us care about those who live in distant lands. We must care no less about what happens to humanity beyond our lifespan. We are buying our good lives at the price of denying good lives to others. Climate change is an intellectual challenge but it is also a test of human solidarity. If we fail that test, we are not worth much: 'No man rises far above the ranks.'

NOTES

(SOURCES OF QUOTATIONS)

Page 10 from: *Answers in Genesis* (AiG). Where Was Man During the Ice Age? By Michael J. Oard, October 1, 2004, paragraph 2.

Page 24 from: Muller, R. A. (2012). The conversion of a climate-change skeptic. *The New York Times* (op-ed), July 28, 2012, paragraph 5.

Page 50 from: Huxley, A. (1928). *Point counter point.* London: Chatto & Windus, page 75.

Page 58 from: Wang Shuo. (1989). *Please don't call me human.* Boston: Cheng & Tsui, page 47.

Page 69 from: About NIF & photon science, web entry 2011 by Lawrence Livermore Laboratories (https://lasers.llnl.gov/about).

Page 77 from: Bennet, J. (2015). We need an energy miracle. *The Atlantic*, November, 2015, paragraph 9.

RECOMMENDED READING

Bentley, C. R., Thomas, R. H. and Velicogna, I. (2007). Ice Sheets. In UNEP (ed.), *Global outlook for ice and snow*, pp. 44–114, Arendal, Norway: UNEP/GRID.

Burroughs, W. J. (2008). *Climate change in prehistory: The end of the reign of chaos*. Cambridge: Cambridge University Press.

Cochran, T. B., Feiveson, H. A., Patterson, W., Pshakin, G., Ramana, M. V., Schneider, M., Suzuki T., & von Hipple, F. (2010). *Fast breeder reactor programs: History and status*. International Panel on Fissile Materials: Research Report 8. Princeton, NJ: International Panel on Fissile Materials.

Davis, S. J., Cao, L., Caldeira, K., & Hoffert, M. I. (2013). Rethinking wedges. *Environmental Research Letters* 8: 011001.

Diamond, J. (2005). *Collapse: How societies choose to fail or succeed*. New York: Viking Press.

Fagan, B. (2002). *The Little Ice Age: How climate made history, 1300–1850*. New York: Basic Books.

Fagan, B. (2008). *The great warming: Climate change and the rise and fall of civilizations*. New York: Bloomsbury.

Ferro, S. (2013). The National Ignition Facility just got way closer to fusion power. *Popular Science*, October 2013.

Gagosian, R. B. (2003). *Abrupt climate change: Should we be worried?* Woods Hole MA: Woods Hole Oceanographic Institution.

Goodall, C. (2010). *How to live a low-carbon life: The individual's guide to tackling climate change*. London: Routledge.

Johnson, E. (2013). Climate engineering might be too risky. *Horizon (The EU Research and Innovation Magazine)*, July 3, 2013.

Kaiser, J. (2001). The other global pollutant: Nitrogen proves tough to curb. *Science* 294: 1268–1269.

Kellow, A. (2007). *Science and public policy: The virtuous corruption of virtual environmental science.* Cheltenham, UK: Edward Elgar.

Kerr, R. A. (2011). Antarctic ice's future still mired in its murky past. *Science* 333: 401.

Mackay, A. W., Battarbee, R. W., Birks, H. J. B., & Oldfield, F. (eds) (2003). *Global change in the Holocene*. London: Arnold.

McCarthy, M. (2011). Global warming's winners and losers. *New Zealand Herald*, December 7, 2011.

Myhrvold, N. P. & Caldeira, K. (2012). Greenhouse gases, climate change and the transition from coal to low-carbon electricity. *Environmental Research Letters* 7: 014019.

NASA (2012). How much more will earth warm? *NASA Earth Observatory*, http://earthobservatory.nasa.gov/Features/GlobalWarming/page5.php. Greenbelt, MA: NASA Goddard Space Flight Center, EOS Project Science Office.

Pielke, R. A. Jr. (2010). *The climate fix: What scientists and politicians won't tell you about global warming.* New York: Basic Books.

Riebek, H. (2011). The carbon cycle. *NASA Earth Observatory*, http://earthobservatory.nasa.gov/Features/CarbonCycle/. Greenbelt, MA: NASA Goddard Space Flight Center, EOS Project Science Office.

Roberts, N. (1998). *The Holocene: An environmental history* (2nd ed.). Malden MA: Blackwell.

Rockstrum et al. (2009). A safe operating space for humanity. *Nature* 461: 472–475. [See the item in the References for the full list of distinguished authors.]

Salter, S., Sortino, G., & Latham, J. (2008). Sea-going hardware for the cloud albedo method of reversing global warming. *Philosophical Transactions of the Royal Society* 366: 3989–4006.

Seaver, K. (1996). *The frozen echo*. Palo Alto, CAL: Stanford University Press.

Soil Carbon Center (2011). *What is the carbon cycle?* Manhattan, KS: Kansas State University.

Smith, L. C. (2010). *The world in 2050: Four forces shaping civilization's northern future.* New York: Plume.

REFERENCES

Angel, R. (2006). Feasibility of cooling the Earth with a cloud of small spacecraft near the inner Lagrange point (L1). *Proceeding of the National Academy of Sciences* 103: 17184–17189.

Anand, R. K., Tallarek, U., & Crooks, R. M. (2013). Electrochemically mediated seawater desalination. *Angewandte Chemie International Edition* 52: 8107–8110.

Anseeuw, W., Alden, W. L., Cotula, L., & Taylor, M. (2012). *Land rights and the rush for land: Findings of the global commercial pressures on land.* Rome: International Land Coalition (ILC).

Australian Bureau of Meteorology. (2016). Australian climate variability and change – Time series graphs. Retrieved 10 September 2013 from www.bom.gov.au/climate/change

Bala, G., Caldeira, K., Nemani, R., Cao, L., Ban-Weiss, G., & Shin, H. (2011). Albedo enhancement of marine clouds to counteract global warming: Impacts on the hydrological cycle. *Climate Dynamics* 37: 915–931.

Ban-Weiss, G. A., & Caldeira, K., (2010). Geoengineering as an optimization problem. *Environmental Research Letters* 5: 1–9.

Barnola, J.-M., Anklin, M., Porchcron, J., Raynaud, D., Schwander, J., & Stauffer, B. (1995). CO_2 evolution during the last millennium as recorded by Antarctic and Greenland ice. *Tellus* 47: 264–272.

Beck, E.-G. (2007). 180 years of atmospheric gas analysis CO_2 by chemical methods. *Energy and Environment* 18: 259–282.

Bennet, J. (2015). We need an energy miracle. *The Atlantic*, November, 2015.

Brennan, D. S. (2013). Sea spray study gains $20 million grant. *UT San Diego*, September 9, 2013.

Bradley, R. S., & Jones, P. D. (1992). Records of explosive volcanic eruptions over the last 500 years. In Bradley, R. S., & Jones, P. D. (eds.), *Climate change since A.D. 1500*, pp. 606–622, London: Routledge.

British Antarctic Survey (2013). Ice cores and climate change. Science Briefing. Retrieved 10 September 2013 from https://www.bas.ac.uk/data/our-data/publication/ice-cores-and-climate-change/.

British Petroleum (June 2011). *BP statistical review of world energy*. Retrieved 8 May 2013 from: http://www.bp.com/en/global/corporate/about-bp/energy-economics/statistical-review-of-world-energy-2013.html.

British Petroleum (June 2013). *BP statistical review of world energy*. Retrieved 8 May 2013 from: http://www.bp.com/en/global/corporate/about-bp/energy-economics/statistical-review-of-world-energy-2013.html.

Chapman, R. (2015). *Time of useful consciousness: Acting urgently on climate change.* Wellington, New Zealand: Bridget William Books.

Christiansen, B., & Ljungqvist, F. C. (2011). The extra-tropical NH temperature in the last two millennia: reconstructions of low-frequency variability. *Climate of the Past Discussions* 7: 3991–4035.

Dansgaard, W., Johnsen, S., Clausen, H. B., Dahl-Jensen, D., Gundestrup, N., Hammer, C. U., & Oeschger, H. (1984). North Atlantic climatic oscillations revealed by deep Greenland ice cores. In Hansen, J. E., Takahashi, T. (eds.), *Climate processes and climate sensitivity*, pp. 288–298, Washington D.C.: American Geophysical Union.

Davis, S. J., Cao, L., Caldeira, K., & Hoffert, M. I. (2013). Rethinking wedges. *Environmental Research Letters* 8: 011001.

deMenocal, P. B. (2001). Cultural responses to climate change during the late Holocene. *Science* 292: 667–673.

Diamond, J. (2005). *Collapse: How societies choose to fail or succeed.* New York: Viking Press.

Dupont, S., Havenhand, J., Thorndyke, W., Peck, L., & Thorndyke, M. (2008). Near-future level of CO_2-driven ocean acidification radically affects larval survival and development in the brittlestar *Ophiothrix fragilis*. *Marine Ecology Progress Series* 373: 285–294.

Etheridge, D. M., Steele, L. P., Langenfelds, R. L., Francey, R. J., Barnola, J.-M., & Morgan, V. I. (1996). Natural and anthropogenic changes in atmospheric CO_2 over the last 1000 years from air in Antarctic ice and firn. *Journal of Geophysical Research* 101: 4115–4128.

Fagan, B. (2002). *The Little Ice Age: How climate made history, 1300–1850.* New York: Basic Books.

Fagan, B. (2008). *The great warming: Climate change and the rise and fall of civilizations.* New York: Bloomsbury.

Ferro, S. (2013). The National Ignition Facility just got way closer to fusion power. *Popular Science*, October 2013.

Flannery, T. F. (2015). *Atmosphere of hope: Searching for solutions to the climate crisis.* Melbourne, Australia: Text Publishing.

Foley, J. A., Asner, G. P., Costa, M. H., Coe, M. T., DeFries, R., Gibbs, H. K., Howard, E. A., Olson, S., Patz, J., Ramankutty, N., & Snyder, P. (2007). Amazonian revealed: Forest degradation and loss of ecosystem goods and services in the Amazon Basin. *Frontiers in Ecology and Environment* 5: 25–32.

Gagosian, R. B. (2003). *Abrupt climate change: Should we be worried?* Woods Hole MA: Woods Hole Oceanographic Institution.

Gillet, N., & Thompson, D. W. J. (2003). Simulation of recent Southern Hemisphere climate change. *Science* 302: 273–275.

Gillis, J. (2014). Looks like rain again. And again. *New York Times*, May 12, 2014.

Goodall, C. (2011). The big biochar experiment. Carbon Commentary, *The Guardian*, October 5, 2011.

Google © (2016). Flood maps: Map of New Zealand, sea-level rise +60 meters. Data from NASA. Viewed 9 June 2016 from http://flood.firetree.net/?ll=-41.1803,175.3917&m=60).

Gorman, S. (2012). The mad, mad world of climatism: *Mankind and climate change mania.* New Lenox, IL: New Lenox Books.

Grove, J. M. (1988). *The little ice age.* London: Methuen.

Handoh, I. C., & Lenton, T. M. (2003). Periodic mid-Cretaceous oceanic anoxic events linked by oscillations of the phosphorus and oxygen biogeochemical cycles. *Global Biogeochemical Cycles* 17: 1092–1103.

Hawkins, E., Anderson, B., Diffenbaugh, N., Mahlstein, I., Betts, R., Hegerl, G., Joshi, M., Knutti, R., McNeall, D., Solomon, S., Sutton, S., Sutton, R., Syktus,J., & Vecchi, G. (2014). Uncertainties in the timing of unprecedented climates. *Nature* 511, http://dx.doi.org/10.1038/nature13523.

Hoffman, G. (2007). Beck back to the future. *Real Climate: Climate Science from Climate Scientists*, May 1, 2007.

Hofmann, G. E., Barry, J. P., Edmunds, P. J., Gates, R. D., Hutchins, D. A., Klinger, T., & Sewell, M. A. (2010). The effect of ocean acidification on calcifying organisms in marine ecosystems: An organism-to-ecosystem perspective. *The Annual Review of Ecology, Evolution, and Systematics* 41:127–147.

Honisch, B., Ridgwell, A., Schmidt, D.N., Thomas, E., Gibbs, S.J., Sluijs, A., Zeebe, R., Kump, L., Martindale, R.C., Greene, S.E., Kiessling, W., Ries, J., Zachos, J.C., Royer, D.L., Barker, S., Marchitto, T.M. Jr., Moyer, R., Pelejero, C., Ziveri, P., Foster, G.L., Williams, B. (2012). The geological record of ocean acidification. *Science* 335: 1058–1063.

Huang, S. P., Pollack, H. N., & Shen, P.-Y. (2008). A late Quaternary climate reconstruction based on borehole heat flux data, borehole temperature data, and the instrumental record. *Geophysical Research Letters* 35: L13703 (5 pages).

Huxley, A. (1928). *Point counter point.* London: Chatto & Windus.

International Energy Agency (2009). *CO_2 capture and storage: A key carbon abatement problem.* Paris: Energy Technology Office, IEA.

IPCC – International Panel on Climate Change (2001). *Climate change 2001: Synthesis Report.* Available at https://www.ipcc.ch/report/ar5/syr/.

IPCC – International Panel on Climate Change (2014). *Climate change 2014: Mitigation of climate change.* Available at https://www.ipcc.ch/report/ar5/wg3/.

Johnson, E. (2013). Climate engineering might be too risky. *Horizon (The EU Research and Innovation Magazine)*, July 3, 2013.

Jouzel, J., Masson-Delmotte, V., Cattani, O., Dreyfus, G., Falourd, S., Hoffmann, G., Minster, B., Nouet, J., Barnola, J. M., Chappellaz, J., Fischer, H., Gallet, J. C., Johnsen, S., Leuenberger, M., Loulergue, L., Luethi, D., Oerter, H., Parrenin, F., Raisbeck, G., Raynaud, D., Schilt, A., Schwander, J., Selmo, E., Souchez, R., Spahni, R., Stauffer, B., Steffensen, J. P., Stenni, B., Stocker, T. F., Tison, J. L., Werner, M., & Wolff, E. M. (2007). Orbital and millennial Antarctic climate variability over the past 800,000 years. *Science* 317: 793–797.

Kaiser, J. (2001). The other global pollutant: Nitrogen proves tough to curb. *Science* 294: 1268–1269.

Keeling, C. D., & Whorf, T. P. (2000). The 1,800-year oceanic tidal cycle: A possible cause of rapid climate change. *Proceedings of the National Academy of Sciences of the United States of America* 97: 3814–3819.

Keenan, R. J., Reams, G. A., Achard, F., de Freitas, J. V., Grainger, A., & Lindquist, E. (2015). Dynamics of global forest area: Results from the FAO Global Forest Resources Assessment 2015. *Forest Ecology and Management* 352: 9–20.

Kellow, A. (2007). *Science and public policy: The virtuous corruption of virtual environmental science.* Cheltenham, UK: Edward Elgar.

Kharecha, P. A., & Hansen, J. E. (2008). Implications of 'peak oil' for atmospheric CO_2 and climate. *Global Biogeochem Cycles* 22: GB3012.

Lamb, H. (1990). As given in W.J. M. Tegart, G.W. Sheldon, & D.C. Griffiths (eds.), *Climate Change: The IPCC Impacts Assessment.* Canberra: Australian Government Publishing Service (Figure 7.1c).

Leach, H. (1984). One thousand years of gardening in New Zealand. Wellington, New Zealand: A. H. & A. W. Reed.

Lovelock, J. (1979). *Gaia: A new look at life on earth.* Oxford UK: Oxford University Press.

Lovelock, J. (2014). *A rough ride to the future.* London: Penguin.

Lund, D. C., Lynch-Stieglitz, J., & Curry, W. B. (2006). Gulf Stream density structure and transport during the last millennium. *Nature* 444: 601–604.

Lüthi, D., Le Floch, M., Bereiter, B., Blunier, T., Barnola, J.-M., Siegenthaler, U., Raynaud, D., Jouzel, J., Fischer, H., Kawamura, K., & Stocker, T. F. (2008). High-resolution carbon dioxide concentration record 650,000–800,000 years before present. *Nature* 453: 379–382.

Manley, S. (2012). Carbon Gold: Working with Cacao farmers in Belize to create a rotating biochar production and utilization system. *International Biochar Initiative,* May 12, 2012.

Mann, M. E., Bradley, R. S., & Hughes, M. K. (1999). Northern hemisphere temperatures during the past millennium: Inferences, uncertainties, and limitations. *Geophysical Research Letters* 26: 759–762.

Matthews. H. D., & Caldeira, K. (2007). Transient climate–carbon simulations of planetary geoengineering. *Proceedings of the National Academy of Sciences* 104: 9949–9954.

McCarthy, M. (2011). Global warming's winners and losers. *New Zealand Herald,* December 7, 2011.

Meyers, S. S., Zanobetti, A., Kloog, I., Huybers, P., Leakey, A. D. B., Bloom, A. J., Carlisle, E., Dietterich, L. H., Fitzgerald, G., Hasegawa, T., Holbrook, N.M., Nelson,

R. L., Ottman, M. J., Raboy, V., Sakai, H., Sartor, K. A., Schwartz, J., Seneweera, S., Tausz, M., & Usui. Y. (2014). Increasing CO_2 threatens human nutrition. *Nature* 510: 139–142.

Mora, C., Frazier, A. G., Longman, R. J., Dacks, R. S., Walton, M. M., Tong, E. J., Sanchez, J. J., Kaiser, L. R., Stender, Y. O., Anderson, J. M., Ambrosino, C. M., Fernandez-Silva, I., Giuseffi, L. M., & Giambelluca, T. W. (2013). The projected timing of climate departure from recent variability. *Nature* 502: 183–187.

Mora, C., Frazier, A. G., Longman, R. J., Dacks, R. S., Walton, M. M., Tong, E. J., Sanchez, J. J., Kaiser, L. R., Stender, Y. O., Anderson, J. M., Ambrosino, C. M., Fernandez-Silva, I., Giuseffi, L. M., & Giambelluca, T. W. (2014). Mora et al. reply. *Nature* 511, http://dx.doi.org/10.1038/nature13523.

Moses, E. (2011). *The National Ignition Facility and the goal of near term laser fusion energy.* Paper delivered at EQEQ 2011: European Quantum Electronics Conference on lasers and electro-optics, Munich, May 22, 2011.

Myhrvold, N. P., & Caldeira, K. (2012). Greenhouse gases, climate change and the transition from coal to low-carbon electricity. *Environmental Research Letters* 7: 014019.

NCCOS – National Centers for Coastal Ocean Science (2014). Average 2014 Gulf of Mexico 'dead zone' confirms NOAA-supported forecast. *NCCOS News and Features.* Retreived 4 August 2014 from https://coastalscience.noaa.gov/news/topics/misc/2014-gulf-of-mexico-dead-zone-large-but-average-confirms-forecast/. Data source: Nancy N. Rabalais & R. Eugene Turner, Louisiana State University.

Neftel, A., Friedli, H., Moor, E., Lötscher, H., Oeschger, H., Siegenthaler, U., & Stauffer, B. (1994). Historical CO_2 record from the Siple Station ice core. In Boden, T. A., Kaiser, D. P., Sepanski, R. J., & Ross, F. W. (eds), *Trends: A Compendium of Data on Global Change*, pp. 11–14. Oak Ridge, TN: Oak Ridge National Laboratory.

New York Times (2012–2016). What could disappear (US coastal levels at 25 feet)? *New York Times* online. Updated 24 April 2016 and retrieved from: http://www.nytimes.com/interactive/2012/11/24/opinion/sunday/what-could-disappear.html?_r=1&

New Zealand Ministry for the Environment (n.d.). Climate change: physical impacts and adaptations. Accessed 7 June 2016 from: http://www.climatechange.govt.nz/.

Nicholls, R. J., & Cazenave, A. (2010). Sea-level rise and its impact on coastal zones. *Science Magazine* 328: 1517–1520.

Nunn, P. D. (2007). *Climate, environment and society in the Pacific during the last millennium.* Amsterdam: Elsevier.

Nye, J. (2013). Apocalypse now: Unstoppable man-made climate change will become reality by the end of the decade and could make New York, London and Paris uninhabitable within 45 years, claims new study. *Daily Mail online*, 10 October 2013.

N'Yeurt, A. D., Chynoweth, D. P., Capron, M. E., Stewart, J. R., & Hasan, M. A. (2012). Negative carbon via Ocean Afforestation. *Process Safety and Environmental Protection* 90: 467–474.

Oard, M. J. (2004). Where was man during the ice age? *Answers in Genesis*. Accessed 7 June 2016 from: https://answersingenesis.org/answers/books/frozen-in-time/where-was-man-during-the-ice-age/.

Orr, J. C., Fabry, V. J., Aumont, O., Bopp, L., Doney, S. C., Feely, R. A., Gnanadesikan, A., Gruber, N., Ishida, A., Foos, F., Key, R. M., Lindsay, K., Maier-Reimer, E., Matear, R., Monfray, P., Mouchet, A., Najjar, R. G., Plattner, G., Rodgers, K. B., Sabine, C. L., Sarmiento, J. L., Schlitzer, R., Slater, R. D., Totterdell, I. J., Weirig, M., Yamanaka, Y., & Yool, A. (2005). Anthropogenic ocean acidification over the twenty-first century and its impact on calcifying organisms. *Nature*, 437, (7059), 681–686.

Owen, J. (2005). Farming claims almost half earth's land, new maps show. *National Geographic News*, December 9, 2005.

Pandolfi, J. M., Connolly, S. R., Marshall, D. J., & Cohen, A. L. (2011). Projecting coral reef futures under global warming and ocean acidification. *Science* 333: 418–422.

Pearce, F. (2016). Hello, cool world. *New Scientist*, February 20, 2016, pp. 30–33.

Philological Society, 1808. The European magazine, and London review: Volume 54. London: Philological Society.

Pielke, R. A. Jr. (2010). *The climate fix: What scientists and politicians won't tell you about global warming.* New York: Basic Books.

Quaile-Kersken, I. (2016). Arctic sea ice set for record summer low. National Snow and Ice Data Center (21 April 2016).

Riebeek, H. (2011). The carbon cycle. *Earth Observatory*, EOS Project Science Office, NASA Goddard Space Flight Center.

Rignot, E., Mouginot, J., Morlighem, M., Seroussi, H. & Scheuchl, B. (2014). Widespread, rapid grounding line retreat of Pine Island, Thwaites, Smith and Kohler glaciers, West Antarctica from 1992 to 2011. *Geophysical Research Letters* http://dx.doi.org/10.1002/2014GL060140

Rockstrom, J., Steffen, W., Noone, K., Persson, A., Chapin, F. S., Lambin, E. F., Lenton, T, M., Scheffer, M., Folke, C., Schellnhuber, H. J., Nykvist, H. J. B., de Wit, C. A., Hughes, T., van der Leeuw, S., Rodhe, H., Sörlin, S., Snyder, P. K., Costanza,

R., Svedin, U., Falkenmark, M., Karlberg. L., Corell, R. W., Fabry, V. J., Hansen, J., Walker, B., Liverman, D., Richardson, K., Crutzen, P., & Foley, J. A. (2009a). A safe operating space for humanity. *Nature* 461: 472–475.

Rockstrom, J., Steffen, W., Noone, K., Persson, A., Chapin, F. S., Lambin, E. F., Lenton, T, M., Scheffer, M., Folke, C., Schellnhuber, H. J., Nykvist, H. J. B., de Wit, C. A., Hughes, T., van der Leeuw, S., Rodhe, H., Sörlin, S., Snyder, P. K., Costanza, R., Svedin, U., Falkenmark, M., Karlberg. L., Corell, R. W., Fabry, V. J., Hansen, J., Walker, B., Liverman, D., Richardson, K., Crutzen, P., & Foley, J. A. (2009b). Planetary boundaries: Exploring the safe operating space for humanity. *Ecology and Society* 14: 32–63.

Russell, J. C. (1972). Population in Europe. In C. M. Cipolla (ed.), *The Fontana Economic History of Europe*, Volume I: *The Middle Ages*, pp. 25–71, Glasgow: Collins/ Fontana.

Salinger, M. J. (1976). New Zealand temperatures since 1300 AD. *Nature* 260: 310–311.

Salter, S., Sortino, G., & Latham, J. (2009). Sea-going hardware for the cloud albedo method of reversing global warming. *Philosophical Transactions of the Royal Society* 366: 3989–4006.

Sample, I. (2014). Sustainable nuclear fusion breakthrough raises hopes for ultimate green energy. *The Guardian*, 13 February 2014.

Schaefer, K., Zhang, T., Bruhwiler, L., & Barrett, A. P. (2011). Amount and timing of permafrost carbon release in response to climate warming. *Tellus B* 63: 165–180.

Schuur, E. A. G., Bockheim, J., Canadell, J. P., Euskirchen, E., Field, C. B., Goryachkin, S. V., Hagemann, S., Kuhry, P., Lafleur, P. M., Lee, H., Mazhitova, G., Nelson, F. E., Rinke, A.,Vladimir, A., Romanovsky, E., Shiklomanov, I.A., Tarnocai, N., Venevsky, S., Vogel, J. G., & Zimov, S. A. (2008). Vulnerability of permafrost carbon to climate change: Implications for the global carbon cycle. *BioScience* 58: 701–714.

Shackley, S., & Sohi, S., (eds.) (2010). *An assessment of the benefits and issues associated with the application of biochar to soil: A report commissioned by the United Kingdom Department for Environment, Food and Rural Affairs, and Department of Energy and Climate Change*. Edinburgh: UK Biochar Research Centre.

Siegenthaler, U., Friedli, H., Loetscher, H., Moor, E., Neftel, A. et al. (1988). Stable-isotope ratios and concentration of CO_2 in air from polar ice cores. *Annals of Glaciology* 10: 151–156.

Smalley, E. (2007). Climate engineering is doable, as long as we never stop. *Wired*, July 25, 2007.

Soil Carbon Center (2011). *What is the carbon cycle?* Manhattan, KS: Kansas State University.

Thompson, D. W. J., & Solomon, S. (2002). Interpretation of recent Southern Hemisphere climate change. *Science* 296: 895–899.

Thorson, R. M. (2014). Politics catches up with climate change. *Hartford Courant*, Thursday, May 15, 2014.

Tilmes, S., Müller, R., & Salawitch, R. (2008). The sensitivity of polar ozone depletion to proposed geoengineering schemes. *Science* 320: 1201–1204.

Turner, J., Comiso, J. C., Marshall, G. J., Lachlan-Cope, T. A., Bracegirdle, T., Maksym, T., Meredith, M. P., Wang, Z., & Orr, A. (2009). Non-annular atmospheric circulation change induced by stratospheric ozone depletion and its role in the recent increase of Antarctic sea ice extent. *Geophysical Research Letters* 36: L08502, 5 pages.

UK Farmers Weekly (2012). Farmers Weekly interactive fertilizer prices. *UK Farmers Weekly*.

Van Bennekom, A. J., Gieskes, W. W. C, & Tisssen, S. B. (1975). Eutrophication of Dutch coastal waters. *Proceedings of The Royal Society B, the Biological Sciences* 189: 359–374.

von Hipple, F. (2010). Overview: The rise and fall of plutonium breeder reactors. In Cochran, T.B., Feiveson, H. A., Patterson, W., Pshakin, G., Ramana, M.V., Schneider, M., Suzuki, T. & von Hippel, F., *Fast breeder reactor programs: History and status*, pp. 1–15, International Panel on Fissile Materials: Research Report 8.

World Nuclear Association (2015). Energy analysis of power systems, updated February 2015.

Zhang, J. (2007). Increasing Antarctic sea ice under warming atmospheric and oceanic conditions. *Journal of Climate* 20: 2515–2529.

Zhang, M., Guo, X., Ma, W., Zhang, S., Huo, L., Ade, H., & Hou, J. (2014). An easy and effective method to modulate molecular energy level of the polymer based on benzodithiophene for the application in polymer solar cells. *Advanced Materials* 26: 2089–2095.

NAME INDEX

Only first authors have been indexed. However, whenever an author has been cited in the text and listed in the references, the index gives page citations for them in both. This allows all sources cited in the text to be identified. The references also contain a few authors not cited in the text but whose works underlie assertions made therein.

SUBJECT INDEX

ALSO BY JIM FLYNN

Philosophy
Humanism and Ideology: An Aristotelian view (1973)
How to Defend Humane Ideals: Substitutes for objectivity (2000)
Fate and Philosophy: A journey through life's great questions (2012)

The modern world
The Torchlight List: Around the world in 200 books (2010)
How to Improve Your Mind: Twenty keys to unlock the modern world (2012)
The New Torchlight List: In search of the best modern authors (2016)

Intelligence
What Is Intelligence? Beyond the Flynn Effect (2007)
Are We Getting Smarter? Rising IQ in the twenty-first century (2012)
Intelligence and Human Progress: The story of what was hidden in our genes (2013)
Does Your Family Make You Smarter? Nature, nurture, and human autonomy (2016)

Group differences
Race, IQ, and Jensen (1980)
Asian Americans: Achievement beyond IQ (1991)

American politics and foreign policy
American Politics: A radical view (1967)
Where Have All the Liberals Gone? Race, class, and ideals in America (2008)
Beyond Patriotism: From Truman to Obama (2012)

Belles lettres
O God Who has a Russian Soul: Poems about New Zealand and its people (2010)

First published in 2016 by Potton & Burton
98 Vickerman Street, PO Box 5128, Nelson, New Zealand
pottonandburton.co.nz

Edited by Sue Hallas

Printed in New Zealand by printing.com

ISBN 978 0 947503 24 6